[英国] 马丁·雷德芬 著 马睿 译

牛津通识读本·

地球

The Earth

A Very Short Introduction

译林出版社

图书在版编目(CIP)数据

地球／（英）雷德芬（Redfern, M.）著；马睿译. —南京：译林出版社，2015.12（2018.4重印）
（牛津通识读本）
书名原文：The Earth: A Very Short Introduction
ISBN 978-7-5447-5778-2

I.①地… II.①雷… ②马… III.①地球－研究 IV.①P183

中国版本图书馆CIP数据核字（2015）第221255号

Copyright © Martin Redfern, 2003
The Earth was originally published in English in 2003. This Bilingual Edition is published by arrangement with Oxford University Press and is for sale in the People's Republic of China only, excluding Hong Kong SAR, Macau SAR and Taiwan, and may not be bought for export therefrom. Chinese and English edition copyright © 2015 by Yilin Press, Ltd.

著作权合同登记号　图字：10-2011-272号

地球　[英国] 马丁·雷德芬　/ 著　马　睿 / 译

责任编辑	何本国
责任印制	董　虎

原文出版	Oxford University Press, 2003
出版发行	译林出版社
地　　址	南京市湖南路1号A楼
邮　　箱	yilin@yilin.com
网　　址	www.yilin.com
市场热线	025-86633278
排　　版	南京展望文化发展有限公司
印　　刷	江苏凤凰通达印刷有限公司
开　　本	635毫米×889毫米 1/16
印　　张	18.5
插　　页	4
版　　次	2015年12月第1版　2018年4月第3次印刷
书　　号	ISBN 978-7-5447-5778-2
定　　价	25.00元

版权所有·侵权必究

译林版图书若有印装错误可向出版社调换　质量热线：025-83658316

序言

陈骏

译林出版社最近出版了一套"牛津通识读本"(*Very Short Introductions*),希望我为其中的《地球》(*The Earth*)一书作序。该书作者马丁·雷德芬(Martin Redfern)毕业于伦敦大学学院地质学专业,是英国广播公司(BBC)科学组资深制作人和蜚声学界的科普作家,先后著有《地心旅行》《瞭望太空》和《行星地球》等。我曾经读过他的作品,对他印象比较深,因而欣然答应。

拿到书之后我很快就将其读完,一个清晰的感受是这是一本好书。要在短短几万字的篇幅中写好一个星球,而且是作为我们所有人栖身家园的星球,堪称一件颇为棘手的工作。它对写作者的挑战,在于不仅要有扎实的专业知识,还要有一定的叙述技巧,既在科学上严谨可信,又让人读出趣味来。马丁·雷德芬无疑不辱使命。这本小书为我们的星球绘制了一幅精彩的肖像。这幅肖像,一如全书开篇的那张地球照片,视野宏大,呈现出地球风起云涌的万千气象;同时又有足够的"像素",放大来看,细节处栩栩如生。读罢掩卷,的确让人有"心事浩茫连广宇"的感觉。

在写作方式上,作者采用了地球系统科学的视角,既把地球看成一个整体,又把它看成动态的系统。如他所说:"在很大程度上,我们像蚂蚁一样在地表忙忙碌碌,对宏阔的图景不知不觉。"正是明确意识到了这一点,作者在写作时着力跳脱出来,仿佛将自己置身于离地球有足够距离的太空之中,终能一窥地球全貌。对这种全景式壮阔景象的呈现,既表现在空间上,也表现在时间上。

比如在介绍"深时"概念时,作者从普通人的感受,即与父母、祖父母、曾祖父母渐次遥远的距离写起,穿越伊丽莎白一世的英格兰、罗马帝国的全盛期、古埃及的大金字塔,仿佛搭乘着一部时间机器,一路写到100亿年前太阳和太阳系尚未诞生之时。沧桑古今,万物流变,时间的深邃幽远跃然纸上。在空间上,书中同样大处着眼,让人感受到的不仅是视野的辽阔,还有基于专业积累之上的想象力。磁层和大气层、生物圈和水圈,以及地球的其他圈层,它们在作者的眼里聚合起来,宛如一颗巨大的洋葱。在亿万年久远的未来,世界地图会有怎样的变化?书中认为,按目前的趋势,"大西洋会继续加宽,太平洋会收缩","澳大利亚会继续北移,赶上婆罗洲,继而转个圈撞上中国"。

为了使叙述更为形象,书中采用了许多生动的比喻。上面提到的"洋葱"即是一例。再比如,在描述大气层时,臭氧被比喻成地球的高效防晒霜。从中世纪教堂的彩色玻璃窗中,可以窥见地幔中硅酸盐岩石的流动方式。由散布于全球各地的测震仪组成的网络,被比喻为身体检查时环绕着患者周身的X射线源和传感器,地震层析成像于是被比拟为我们所熟悉的CT(计算机辅助断层扫描)。再比如,海洋中的洋脊像是网球的接缝、地球的演化一如橘子果酱的制作过程,火山的突然喷发又好似充分摇动香槟之后拔去瓶塞——尽管在时空尺度上,它们被极大地缩小了。这些比喻不仅能帮助读者更好地理解那些远离日常生活的陌生过程,更使阅读充满乐趣。

乍看书名,读者可能认为,一本关于地球的书与我们的日常生活相去甚远,它提供的只是满足好奇心的知识,并无多少"实用"价值。读到书末,尤其是关于地震的那一章,这种印象无疑会有所改变。地震不可避免,但尽力减少地震造成的伤害无疑是值得探索的。在分析伤亡原因时,作者认为倒塌的建筑物和后续的火灾是主因。从建筑材料和设计来看,柔韧的材料比脆硬的要抗震,好的设计(如避免与地震波频率共振、高楼屋顶配备重物抵消晃动)同样必要;在减少火灾伤害方面,旧金山等地震易发城市正在开发"智能管道"系统,无疑是值得借鉴的做法。

特别是,在地震预警方面,书中提供了很好的思路。我们无法预知下一次地震发生的确切时间,但是,发生的概率是可以计算的。从1/36 500,到1/1000,再到1/20,通过综合不同的要素,准确率在逐步逼近;虽然还未达到能向公众播报以疏散人员的程度,但至少可以提供给相关的救援部门,让他们随时待命。通过设置在断层中的传感器,利用无线电的光速与地震波的声速之间的时间差,也会为银行备份、电梯停运、管道封锁等做好准备。这些都是我们可以借鉴的实用经验。

这本小书并不只有科学的视角,它还表现出了作者深切的人文关怀。目前为止,人类尚未发现地球之外的智慧生命,也未发现有其他星球堪为人类的第二家园。遗憾的是,对于这颗蓝色宝石,我们却不够珍惜。"我们早已不再是这个星球的受害者,而变成了它的托管人。而我们却恩将仇报,对土地粗暴轻率地贪婪,对污染置若罔闻地轻忽。但这样做是要承担风险的。我们仍然别无退路,毕竟所有的人都住在同一个星球上。我们应该照顾好这个星球,为它承担起责任。"爱护地球、保护地球不仅是科学家、政府和少数志愿者的工作,更是每一个合格地球公民的责任。承担起这一责任的前提则是了解地球、认识地球,这应当成为每一个"地球人"的基本素养和必修课,应当成为当今大学通识教育不可或缺的重要组成部分。它并非只是关于地球科学知识的概览和介绍,而更是通过引导普通公众和大学生参与到对地球的科学探索和对事关地球与人类发展的重大问题的讨论之中,培养起人们尊重地球、热爱地球、保护地球的意识,自觉地过一种与周围环境相和谐的健康、"绿色"、文明的生活。

这本《地球》是绝佳的科普和通识教育读物。科普读物需要把握科学性与人文性、专业性与普及性之间的平衡,其难点在于既要不失科学的准确、严谨,同时又能让非专业读者和大学生接受、喜爱。与他以往的《地心旅行》等科幻小说相比,这本书虽然没有小说中那些天马行空的想象和惊险刺激的情节,但是同样跌宕起伏、扣人心弦,带给读者一场紧张兴奋的智力冒险,正如伦敦大学学院比尔·麦圭尔(Bill McGuire)的书评所言:"关于地球的迷人真相呼之欲出。这是一个关于矿物、岩浆和地质灾难的故事,内核飞旋、板块炸裂,无不惊心动魄。"这是一场精心安排的冒险之旅,是对地球主要运作机制的全景扫描,并对重要"景点",诸如地球历史、板块运动、海洋结构、火山地震等一一作了精准解说。阅读本书,是一趟丰富充实的地球知识之旅,也是充满人文内涵和艺术享受的思想之旅、审美之旅。

感谢译林出版社出版这套通识读本。近几年,包括南京大学在内的中国多所著名大学已充分认识到通识教育的重要性。南京大学率先探索本科教育改革,目前正在稳步推进,力图办中国最好的本科教育。通识教育是一个系统工程,需要多方面因素的合力来推动；其中，一套高质量的读本是具有基础性作用的。包括这本《地球》在内的"牛津通识读本"丛书，一定会对我国高等教育改革和提高国民素质提供帮助，祝愿在中国大学的通识教育事业中，在所有渴求知识的人丰富自己的精神世界和人生画卷的道路上，这套书能发挥更大的作用。

目录

致谢 *1*

1 动态的地球 1

2 "深时" 21

3 地球深处 37

4 海洋之下 56

5 漂移的大陆 77

6 火山 99

7 地动山摇之时 116

结语 131

索引 134

英文原文 141

致谢

　　本书作者特此感谢以下诸位：感谢阿琳·朱迪丝·克洛茨科，本书的撰写离不开她的引介；感谢谢利·考克斯当初热心约稿、埃玛·西蒙斯一贯耐心相助、戴维·曼及时提供插图、保利娜·纽曼和保罗·戴维斯提出颇有助益的意见、玛丽安和埃德蒙·雷德芬助我热情饱满并帮我审读书稿、罗宾·雷德芬也卖力相助；感谢激励我始终保持严谨的无名读者，以及拨冗与我交流并用激情将我深深感染的无数地质学家。

图1 1972年12月从"阿波罗"17号上看到的行星地球

第一章
动态的地球

> 一旦有人从外太空拍摄一张地球的照片，一种前所未有但无可辩驳的全新观念就要诞生了。
>
> ——弗雷德·霍伊尔爵士①，1948年

如何在薄薄一本小册子里容纳一个巨大的星球？尺幅千里已显不足，不过倒有两种天差地别的方法可供一试。一种是地质学采用的自下而上的方法：从本质上说，就是观测岩石。数个世纪以来，地质学家们奔波于地球表面，用手中的小锤子探测不同的岩石类型以及构成这些岩石的矿物颗粒。他们先是利用肉眼和显微镜、电子探针和质谱仪，把地壳简化为细小的组分。继而又绘制出不同的岩石类型之间的联系，并通过理论、观察和实验，提出了岩石运动的假说。他们从事的是一项艰巨的事业，提出了不少深刻的见解。地质学家前仆后继的努力构筑了一座理论大厦，为未来的地球科学家奠定了基础。正因为有了这种自下而上的方法，我才得以撰写本书，但这并不是我要采纳的视角。我

---

① 弗雷德·霍伊尔爵士（1915—2001），英国天文学家，英国皇家学会会员。著有多部学术专著、科普读物、科幻小说，以及一部自传。霍伊尔的许多研究成果不符合正统的学术观点，但仍然被视为20世纪最有影响力的科学家之一。

本意并不想写一本岩石矿物和地质制图指南，而是要为一个星球画一幅肖像。

要观察我们这个古老的星球，还有一个自上而下的新方法，也就是日渐为人们所知的地球系统科学的视角。它把地球看成一个整体，而不仅仅是"现在"这一刻凝固的模样。采纳了地质学的"深时"概念，我们开始把这个星球看成一个动态的系统，由一系列过程和循环组成。我们开始了解它的运作机理。

俯瞰

开篇那句话是天文学家弗雷德·霍伊尔爵士在1948年提出的预言，仅仅十年后，人类就开始了太空之旅。当无人驾驶的火箭在外太空拍下第一批地球照片时，当第一代宇航员亲眼看到我们这个世界的全貌时，预言成真。最初的俯瞰并没有揭示什么关于地球的新秘密，却已成为一个充满象征意义的符号。对于亲眼看到那幅图景的许多宇航员来说，那是一次动人的体验：他们一直以来与之共存的这个世界竟然美丽如斯，又显得那么脆弱。地球科学也在同时经历了自身的革命，这大概不是偶然。板块构造的概念最终为世人所接受，其时距离阿尔弗雷德·魏格纳[①]首次提出该理论已逾50载。海底探索揭示出海底是从洋中脊系统扩展而成的。它不断漂移，迫使大陆分离或重组成新大陆。那些大陆一般大小的岩石板块，其形体之巨远超想象，却也翩然跳出绚丽而古老的舞步。

大约同时，与人类在广袤幽暗的太空中俯瞰到"地球"这个

---

① 阿尔弗雷德·魏格纳（1880—1930），德国地质学家、气象学家和天文学家，大陆漂移学说的创立者。1930年11月，他最后一次前往格陵兰探险时身亡，享年50岁。

飘浮于其中的小小蓝宝石一样具有象征意义,全球兴起了一场环保运动,它的参与者既有对濒临灭绝的物种和雨林充满感伤主义眷恋的普通人,也有开始采纳全新视角、研究复杂互动的生态系统的科学家。如今,多数大学院系和研究组织都会用"地球科学"一词取代"地质学",因为人们已经意识到该学科的广度绝不仅限于研究岩石。"地球系统"一词也被日益广泛地使用,因为人们认识到这些过程之间相互关联的动态性质,不仅包括由岩石构成的固态地球,也包括其上的海洋、脆弱的大气层,及其表面的薄薄一层生命体。我们生活的世界仿佛一颗洋葱,由一系列同心圈层构成:最外是磁层和大气层,继而是生物圈和水圈,再到固态地球的多个圈层。它们并非都是球形,有些圈层的实体性也远不如其他,但每个圈层都在努力维持着微妙的平衡。人们认为,这个系统的每个组成部分都不是固定不变的,其样态更像一个喷泉,整体结构或许能够维持不变,却会随着通过的物质和能量的大小而展现出不同的姿态。

如果岩石能开口说话

岩石和石子可算不上口若悬河的说书人。它们静坐着,任凭青苔聚集,推一把才会动一下,生性如此。然而地质学家有许多办法令其开口。他们敲打之、切割之、挤之压之、推之拖之,直到它们开口讲述——有时还真要裂开才行。如果你懂得如何观察,岩石会将它的历史娓娓道来。岩石表面是最近的历史:它如何受到风化侵蚀;那些风、水和冰留下的创痕,是它沧桑的容颜。还有些表面看不到的疤痕记录着热量和压力的时期,以及这块岩石被埋葬时的变形情况。当这些变化较为极端时,会形成所谓的变质岩。关于岩石的来历也不乏线索。有些痕迹表明,它曾被熔融并从地球深层强推而上,在火山爆发时喷薄而出,或者侵入存在于地表的其他岩石,这些是火成岩。岩石内部矿物颗粒的大小可以揭示它们冷却的速度有多快。大块花岗岩冷却缓慢,因此其中的晶体很大。火山玄武岩的固化速度很快,因而颗粒细小。先前的岩石经过碾压的碎片会组成新的岩石。就这些岩石而言,碎片的大小往往能够反映其形成过程中环境的力量:从在静水中沉积而成的细粒页岩和泥岩,到沙岩,再到由汹涌水流冲刷而成的粗粒砾岩。其他岩石,诸如白垩和石灰岩,乃是在生命系统从大气中吸收二氧化碳并使之在海水里迅速凝结的过程中,由化学物质沉积而成的,这一过程听上去仿佛是把蓝天变成了顽石。

就连单个的矿物颗粒也有自己的故事。矿物学家能利用高精度质谱仪逐一分辨这些矿物颗粒的微小成分,甚至能够揭示痕量成分中同位素的不同比例(即同种成分的不同原子的排列方式)。有时这些数据能帮助我们确定矿物颗粒形成的年代,从而了解它们是否来自更加古老的岩石。矿物学家还能揭示某一晶体(如钻石)穿过地幔时经历了哪些阶段。就从海生生物体中提取的矿物质而言,研究其碳和氧同位素,甚至有助于测算在这些矿物质形成时海水的温度和全球气候。

其他世界

地球的问题就在于,我们只有一个地球。我们只能看到它当前的状态,无法判断这一切是不是一场美妙的巧合。这也就是地球科学家把关注的目光重新投向天文学的原因。有些新型望远镜的功能很强大,对红外和次毫米波长辐射极为敏感,能够用于深度观测恒星形成区,了解在我们这个太阳系生成的过程中，曾经发生过怎样的故事。在某些年轻的恒星周围，人们通过望远镜观察到满是尘埃的光环，即所谓的原行星盘，它们有可能是正在形成的新的太阳系。不过要找到一些完整成型的地球类行星则比较困难。直接观察这样一个行星围绕一颗遥远恒星的轨道旋转，就像在高亮度探照灯附近寻找一只小小的飞蛾。然而近年来，人们通过间接方法发现了一些行星，主要方法是监测母恒星在运动过程中由于重力作用产生的微小摆动。作用最明显，因而也最先被发现的，似乎是由于那些行星比木星大得多，它们与其所环绕的恒星之间的距离也远小于地日距离。这样一来，就很难将其定义为"地球类行星"了。不过越来越多的证据表明，宇宙中确乎存在与我们的所在更加相像的多行星太阳系，但要找到像地球这样宜人的小行星可没那么容易。

为了直接看到这样的行星，需要使用人类一直梦寐以求的太空望远镜。美国和欧洲都在实施野心勃勃的计划，力图创建一个红外望远镜网络。其中每一台都要比哈勃太空望远镜大得多，必须将四五台这样的望远镜密集编队，把它们的信号组合在一起，才能解析整个行星。这些望远镜必须安置到木星那么远的位置，才能摆脱我们这个行星系所产生的浑浊不清的红外光的干扰。但那样一来，这些望远镜或许能够探测到遥远行星大气层中的生命迹象，尤其是，它们或许能够探测到臭氧。那也许意味着类地的气候和化学条件，外加游离氧的存在，据我们所知，这是只有生命体才能够维持的物质。

生命的迹象

1990年2月，"旅行者"1号探测器在遭遇木星和土星之后冲出太阳系，途中传回了整个太阳系的第一张图片；如果真有外星来客，他们看到的太阳系大概就是那样一幅图景。太阳这颗耀眼的恒星占据了整个画面，那已经是从60亿公里之外拍摄的，相当于我们通常观察太阳的距离的40倍。从图片上几乎看不到任何行星。地球本身比"旅行者"号携带的相机中的一个像素还要小，它发出的微弱光芒则缥缈如一束日光。这是我们全部的世界，看上去却不过一粒微尘。但对于任何携带着适当工具的外星访客而言，那个小小的蓝色世界会立即引起他们的注意。与外行星狂暴肆虐的巨大气囊、火星的寒冷干燥或金星的酸性蒸汽浴不同，地球的一切条件都恰到好处。这里存在三种水相——液体、冰和蒸汽。大气组成不是已经达到平衡的死寂世界，而是活跃的，必须持续更新。大气中有氧气、臭氧，以及碳氢化合物的痕迹；这些物质如果不是在生命过程中持续更新，就不会长时间共存。这本身足以引起外星访客的注意，更不用说这里还有通讯、广播和电视设备不绝于耳的聒噪了。

磁泡

我们对地球物理学还所知甚少。这并不是说这门学问深不可测，而是我们这颗行星的物理影响大大延伸到星球的固体表面之外，深入我们所以为的寂寥太空。但那里并非虚无。我们住在一串泡泡里，它们像俄罗斯套娃一样层层嵌套。地球的势力范围之外，是由太阳主宰的更大的泡泡。而那个大泡泡之外则是彼此重叠的多个泡泡，它们是很久很久以前由恒星或超新星爆炸产生的碎片不断膨胀而形成的。所有这些泡泡都存在于我们的银河系中，银河系则是已知的宇宙之中诸多星系所组成的超星团的一员,而这个超星团本身,可能只是诸多世界的量子泡沫[①]中的一个泡泡而已。

在大多数情况下,地球的大气层和磁场都在保护我们免受来自太空的辐射危害。如果没有这层保护,地球表面的生命就会受到太阳紫外线和X射线、宇宙射线,以及星系间剧烈事件所产生的高能粒子的威胁。太阳还终年不断地向外吹送粒子风,其组成主要是氢原子核或质子。这股太阳风一般以400公里每秒的速度掠过地球,在太阳暴期间,速度会增加三倍。它会弥漫到数十亿公里之外的太空中,越过所有行星,也许还会越过彗星的轨道,那些彗星轨道与太阳的距离要比地日距离远上数千倍。太阳风非常稀薄,但足以在彗星接近太阳系的心脏部位时吹散彗尾,因此,彗尾总是指向偏离太阳的一方。在用薄如轻纱的巨型太阳帆来推进航天器这类富有想象力的提议之中,运用的也是同样的原理。

地球凭借着自身的磁场,即磁层,来躲避太阳风。太阳风带电,所以是一种电流,无法穿越磁场线。相反,它会压缩地球磁层的向阳一侧,就像是海上行船时的顶头波,并且顺着风向拖出一个长尾,其长度几乎能够触及月球轨道。磁层中捕获的带电粒子在磁场线之间组成粒子带,并被迫旋转,从而产生了辐射。1958年,詹姆斯·范艾伦[②]在美国"探险者"1号卫星上放置了

[①] 又称时空泡沫,是1955年由美国物理学家约翰·惠勒(1911—2008)提出的量子力学概念。"泡沫"即为概念化的宇宙结构的基础。量子泡沫可用来描述普朗克长度(10^{-35}米级别)的次原子粒子时空乱流。在如此微小的时空尺度上,时空不再平滑,不同的形状会像泡沫一样旋生旋灭。

[②] 詹姆斯·范艾伦(1914—2006),美国太空科学家。他促成了太空磁层场研究的创立,以发现范艾伦辐射带而知名。

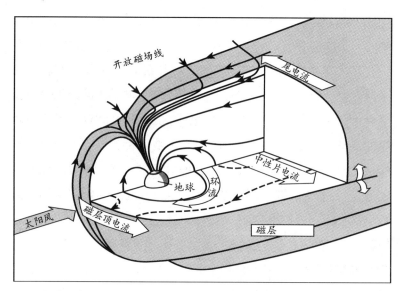

图2 地球磁包层图,太阳风将磁层向后扫进一个彗星式的结构。箭头指示电流的方向

太空中第一个盖革计数器,首次发现了这些辐射带。要想延长航天器的使用寿命,就必须避开这些区域,没有防护设备的宇航员一旦进入这些区域,也会性命难保。

地球的磁场线冲向极地,太阳风中的粒子也会在那里进入大气层,向下射出活跃的原子,产生壮观的极光。在大气层顶部,太阳风自身的氢离子会产生粉红色的薄雾,其下方的氧离子产生红宝石色的辉光,而同温层中的氮离子则产生蓝紫色和红色的极光。偶尔,太阳风中的磁场线会被迫靠近地球的磁场线,使两者重新接合,往往会导致能量的巨大释放,形成更加壮观的极光。

脆弱的面纱

大气层的顶部没有明确的高度；航天飞机所处的近地轨道距地面260公里，那里的气压只有地面的十亿分之一，几乎就处于大气层之外了。但那里每一立方厘米空气中仍有大约十亿个原子，这些热粒子带电，因而对航天器会有腐蚀作用。在太阳活动最剧烈的时期，大气层会轻微膨胀，对近地航天器产生更大的摩擦阻力，所以为了让这些近地航天器保持在轨道之中，必须加大推力。80公里之上的顶部大气层有时被称为热大气层，因为那里非常热，不过那里的空气非常稀薄，不会灼伤皮肤。

大气层的这一区域还会吸收太阳发出的危险的X射线和部分紫外线辐射。因此，很多原子被"离子化"了，也就是说它们会失去一个电子。基于这一原因，热大气层也叫电离层。电离层导电，会反射某些频率的无线电波，这就是为什么全世界的人可以通过设置在地平线上方的发射机传送信号，听到短波无线电的广播。

地面之上仅仅20公里处，在热大气层、中间层和大部分同温层之下，大气中的空气含量仍不足10%。正是在这一高度附近，存在一个稀薄的臭氧层，所谓臭氧，即含有三个氧原子的分子。含有两个原子的普通氧分子被太阳辐射分离时，其中的一些就会重组为三原子的臭氧。对地球而言，臭氧是一种高效的防晒霜。如果地球大气层中所有的臭氧都浓缩在地面，就会形成一层只有约三毫米厚的臭氧层。但它依然能够过滤近乎全部的短波紫外线辐射——这是太阳发出的最危险的辐射——以及大部分中波紫外线。因此，臭氧层让生命免受晒伤和皮肤癌的威胁。由于人类活动所释放的氯氟烃等化学物质的严重破坏，整个臭氧层变得稀薄，而在清寒的春季，极地区域上空特有的臭氧空洞也越来越多。各个国际协定的签署有效放缓了氯氟烃的释放，臭氧层应该会恢复如初，但化学物质会长期存在，臭氧层的复原仍需时间。

循环与周期

对流层位于大气层最接近地面的15公里范围内，是活动最频繁的区域。天气变化就发生在这里。云生云消，风起风止，地球上的温湿转换，都在这里进行。在一个充满生机的星球上，一切看上去都像是能量的循环和流动。在靠近地表的对流层，这些循环都是由太阳能驱动的。随着地球绕自轴旋转，生成明显的昼夜更迭，地面也随之冷热交替；地球绕太阳公转则产生了一年内的季节变换，这是因为两个半球轮换着接收到更多的阳光。但还有比这更长的周期，比如地轴以数十万年为周期来回摆动。

就像地球绕太阳旋转一样，月球也绕着地球旋转。绕行一圈大约需要28天，这正是月份的由来。随着地球绕轴自转，月球的引力拉动地球周围的海水上涨，引发了潮汐。潮汐还能够抑制地球自转，昼夜更迭随之放缓。四亿年前的珊瑚化石上的日增长带表明，当时的昼夜时间要比现在短几个小时。

月球有助于地球的公转轨道保持稳定，从而稳定了气候。不过还有比这长得多的周期在起作用。地球围绕太阳公转的轨道并非标准圆形，而是一个椭圆，太阳位于两个焦点之一。因此，地日距离随着地球的公转而时刻变化着。此外，变化程度本身也会在95 800年的周期内发生变化。而地球的自转轴也像失去平衡的陀螺那样，缓慢摇摆或按岁差旋进。在一个21 700年的时间周期,地轴的轨迹可以描绘成一个完整的锥形。当前,地球在北半球的冬季距离太阳最近。地球自转轴与其绕太阳公转的轨道间的倾斜度(倾角)也会在41 000年的时间周期内发生变化。这些所谓的米兰柯维奇循环[①]经过数万乃至数十万年的累积,就会对气候产生影响。325万年来,地球受到冰川期等现象的影响,都被归咎于米兰柯维奇循环。但事实上原因很可能更加复杂,这些循环的作用往往还会被海洋循环、云量、大气组成、火山气溶胶、岩石的风化、生物生产力等因素放大或缩小。

太阳周期

变化周期并不仅限于地球,太阳也会变化。在其50亿年的生命历程中,太阳变得越来越热。然而同一时期,由于温室气体水平下降,地球的表面温度却要恒定得多。这主要是生命体在起作用,植物和藻类消耗了大量二氧化碳,而二氧化碳的作用就像一张毛毯,给年轻的地球保暖。太阳还发生了其他变化。常规的太阳周期为11年,其间太阳黑子活动由盛转衰,继而反映出太阳磁活动的周期,而太阳磁活动产生了太阳暴和太阳风。其他类日恒星似乎有大约三分之一的时间没有太阳黑子,这一状态被称为蒙德极小期[②]。我们的太阳在公元1645—1715年间曾发生过这种情况。太阳能只下降了大约0.5%,却足以让北欧陷入所谓的"小冰川期",经历一连串极其严酷的冬天。彼时寒冬冰封了伦

---

① 塞尔维亚地球物理学家和天文学家米卢廷·米兰柯维奇在描述地球气候整体运动时所提出的理论。

② 英国天文学家爱德华·沃尔特·蒙德(1851—1928)在进行太阳黑子与太阳磁力周期的研究时,发现在大约1645至1715年的这段时间,太阳黑子非常稀少,这个时期即后来以他的名字命名的蒙德极小期。

敦的泰晤士河，也就有了在冰面上举办的集市和霜降会[①]。

炙热的空气

太阳并非普施温暖，赤道地区最暖和。空气受热膨胀，气压升高。为恢复平衡，就有了风和空气流通。在这一切发生的同时，地球继续自转，空气也因而获得了角动量。赤道地区的角动量最大，结果产生了所谓的科里奥利效应。大气层与固体的星球并非紧密耦合，因此，当赤道风起时，风的动量与其下地表的自转无关。也就是说相对于地表，风在北半球弯向右侧，而在南半球则弯向左侧。这形成了高气压和低气压的空气循环体系，也就是给我们带来雨水或阳光的天气系统。

大片陆地和山脉也会影响热循环和水分循环。比如，在喜马拉雅山脉隆起之前是没有印度季风的。最重要的是，海洋在储存热能和环球传输热能方面发挥了重大作用。海洋顶部两米的热容量与整个大气层相当。与此同时，洋流中也进行着热循环。但表层环流并非全部。北大西洋的湾流就很能说明问题。北大西洋湾流携带着来自墨西哥湾的暖水流向北部和东部，这也是欧洲西北部冬季的天气比美国东北部温和得多的原因之一。随着暖水流向北方，其中一部分蒸发到云中，因而即使英国人外出度假，他们头顶上似乎也总是笼罩着这团裹带着水蒸气的云。余下的海洋表层水冷却下来，盐度也与日俱增。如此一来，这些表层水的密度也会上升，最终下沉，向南流到大西洋深处，大洋环流的传送带至此结束。

[①] 17至19世纪初期的某些冬天在伦敦泰晤士河潮汐水道上举办的集市。

突然袭来的严寒

大约11 000年前,地球结束了最后一次冰川期。冰雪融化,海平面上升,气候普遍变暖。紧接着,不过数年之后,天气突然再次变冷。这一变化在爱尔兰尤为突出,在那里,沉积岩芯中的花粉显示,植被突然从温带疏林恢复成苔原,后者主要是一种仙女木属植物。拉蒙特-多尔蒂地质观测所的沃利·布勒克尔对当时可能的情况进行了研究。随着北美地区的冰原后退,融化的淡水在加拿大中部形成了一个巨大的湖,其规模远比如今的北美五大湖地区要大得多。起初,湖水沿着一条大岩脊的方向流入密西西比河。随着冰层后退,东面突然出现了一条流向圣劳伦斯河的水路,海拔低得多。这个巨大冰冷的淡水湖几乎立即流向了北大西洋。入海的水量如此之大,海平面随即上升了30米之高。入海的淡水稀释了北大西洋表层水的盐分,事实上制止了大洋环流的传送带。因此,再也没有流向北大西洋的暖流,极寒天气卷土重来。1000年后,就像此前突然消失一样,大洋环流突然又重新开始,温暖的气候也回来了。

北大西洋的深水与南极地区冰冷的底层水一起,在远至印度洋和太平洋的深处找到了归宿。深层流持续汇入北太平洋,在其再次上升到表面之前,慢慢积累了各种养分。

全球温室

地球大气层中某些气体的作用就像温室的玻璃一样,把阳光放进来加热地表,却也能阻止所产生的红外热辐射逸出。如果不是温室效应的作用,全球平均气温会比现在低15℃左右,生命几乎无法维系。温室气体主要是二氧化碳,但包括甲烷在内的其他气体也起着重要作用。水蒸气也一样，人们有时会忘记它在这方面的贡献。在数亿年时间里，植物通过光合作用从大气中消除二氧化碳，动物通过呼吸产生二氧化碳，已经形成了一个大致的平衡。大量的碳埋藏在石灰岩、白垩和煤炭等沉积物中，火山爆发则将碳从地球内部释放出来。

近年来，人们越来越关注所谓的温室效应加剧，即人类活动引起大气层中的温室气体水平显著上升。煤炭和石油等化石燃料的燃烧是罪魁祸首，但农业活动会产生甲烷，砍伐森林会从木材和土壤中释放二氧化碳，植被减少使得二氧化碳无法被再次吸收等，也难辞其咎。气候模型显示，这些活动可能会导致全球气温在下个世纪上升若干度，同时伴随着更为剧烈的极端气候，并有可能导致海平面上升。

气候变化

1958年以来，有人仔细记录了夏威夷某座孤峰上二氧化碳水平持续的年增长率。全世界连续130多年的精确气候数据证实，全球平均气温升高了半度左右，最近30年的影响尤为显著。但自然界的气候记录可以追溯到很久以前的远古时期。树木的年轮载列出在它们存活期间干旱和严霜的发生以及野火的频率。从现今保存的木材的重叠层序向前推测，可以显示5万年前的气候状况。珊瑚的生长轮可揭示同一时段的海表温度。沉积物中的花粉粒记录了700万年间植被格局的变迁。地形展现了过去的冰川作用和数十亿年间的海平面变化。但某些最为精确的记录来自钻探得到的冰核与海洋沉积物。冰核不但显示了积雪的速度及圈闭的火山灰，冰里的气泡还提供了雪中圈闭的远古大气样本。氢、碳和氧的同位素也能标示当时的全球温度。如今,南极洲和格陵兰的冰层记录能够追溯到40万年以前。大洋钻井计划在全球各处的海洋沉积物中取样,可获得远至1.8亿年前的记录。圈闭在这些沉积物中的微体化石的同位素比值可以揭示出温度、盐度、大气二氧化碳水平、大洋环流,以及极地冰冠的范围。所有这些不同的记录表明,气候变化是无可逃遁的现实,在漫长的过去,气候要比我们如今体验到的暖和得多。

生命的网络

生命是地球最脆弱的圈层,但它或许对地球产生了最为深远的影响。如果没有生命,地球也许会像金星一样,成为一个失控的温室世界,或是像火星一样,成为一片寒冷的沙漠。当然也不会有温和的气候和为我们提供养分的富氧大气。我们已经知道,原始藻类通过吸收二氧化碳覆盖层,破坏了年轻地球的隔热罩,一度跟为之供暖的太阳亦步亦趋。独立科学家詹姆斯·洛夫洛克认为,这样的反馈机制将陆地气候维持了20亿年以上。他使用古希腊神话中大地女神"盖亚"(Gaia)的名字为该机制命名。洛夫洛克并未假称这一控制中存在着任何有意识或有预谋的成分;"盖亚"没有什么神力。但主要以细菌和藻类形式存在的生命,的确在这一自我平衡的过程中起到了举足轻重的作用,使地球变得宜居。一个名为"雏菊世界"的简单计算机模型显示,两个或两个以上的竞争物种能够在适于生存的限制条件下建立起控制环境的负反馈机制。洛夫洛克猜想,随着人类活动加剧温室效应,地球的全球系统会逐步适应这种变化,即使这类适应可能对人类不利。

碳循环

碳元素在无休无止地移动。每年大约有1280亿吨碳通过陆地上的种种过程，以二氧化碳的形式释放到大气中，而几乎同样大量的碳又会立刻被植物和硅酸盐岩石风化所吸收。海上的情况与之类似，只是吸收量比释放量略多一些。如果没有火山爆发以及每年燃烧化石燃料所释放的50亿吨碳，整个系统大致会处

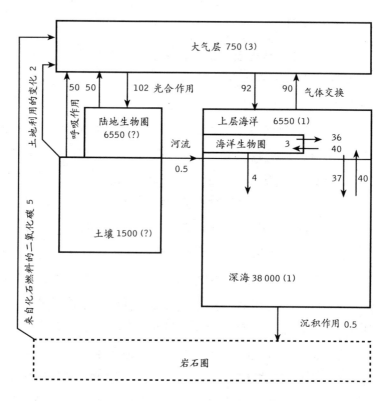

图3 碳循环。这一简化图表显示了储存在大气层、海洋和陆地中的碳总量（以10亿吨为单位）估计值。箭头旁的数字表示储存量的年度变化值，括号中的数字表示年度净增值。与大多数其他变化量相比，燃烧化石燃料的贡献值很小，但足以打破平衡

于平衡状态。大气层中保持的碳总量相当少——只有7.4亿吨，仅比陆地上的动植物所保持的略多一些，比海洋生物所保持的略少一些。相比之下，以溶剂方式储存在海洋中的碳总量洋洋可观，高达340亿吨，而储存在沉积物中的碳总量比这还要高出2000倍。因此，在碳循环中，溶解和沉淀等物理过程可能比生物过程更为重要。但生命体似乎手握王牌。浮游植物所结合的碳会被释放回海水中，如果不是由于桡足动物粪球粒的物理属性，接下来也将非常迅速地释放到大气中。桡足动物是一种微小的浮游动物，其排泄物是质地紧密的小颗粒，可以缓慢地沉入深海，至少暂时将其中所含的碳从碳循环中除去。

宛若洋葱

地球的内部就像一颗洋葱，由一系列同心壳或同心层组成。顶部是一层硬壳，海洋下的平均厚度为7公里，大陆下为35公里。这层硬壳位于地幔上方坚硬的岩石层之上，其下是较为柔软的软流层。上地幔的范围深达670公里左右，下地幔则深至2900公里。在一层薄薄的过渡层下，是熔融的铁所组成的液态外核以及大小相当于月球的固态铁质内核。但这不是一颗完美的洋葱。各圈层之间存在着横向差异，圈层的厚度也不尽相同，而且我们现在知道，圈层之间还持续进行着物质交换。我们的星球与完美的洋葱模型之间最显著的差异，正是现代地球物理学最不关心也最漠视的部分，而我们恰可以从那里找到线索，了解是哪些过程在驱动整个系统的运行。

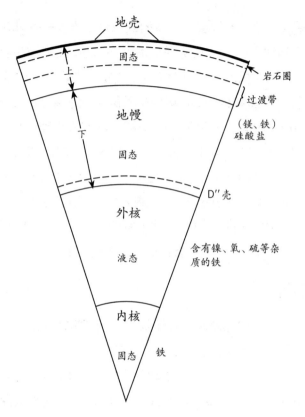

图4 地球径向切面上显示的主要"洋葱"圈层

熔岩灯

还记得1960年代流行的熔岩灯和后来数不清的复古产品吗？它们是地球内部运转过程的绝佳模型。关灯时，在透明的油质层之下有一层红色的黏性半流体物质。开灯时，底部的灯丝将其加热，这层红色半流体受热膨胀，密度因此降低，开始以延展的块状上升到油层顶部。一旦充分冷却，它又下沉回原位。地球的地幔中情况也是如此。放射性衰变和地核所产生的热量驱动

行星数据	
赤道直径	12 756公里
体积	1.084×10^{12}立方公里
质量	5.9742×10^{24}公斤
密度	水的5.52倍
表面重力	9.78米/秒$^{-2}$
逃逸速度	11.18公里/秒$^{-1}$
昼长	23.9345小时
年长	365.256天
轴向倾角	23.44°
年龄	大约46亿年
距离太阳	近日点:1亿4700万公里
	远日点:1亿5200万公里
表面积	50亿9600万平方公里
陆地表面	1亿4800万平方公里
海洋覆盖	71%的表面积
大气层	氮气78%,氧气21%
陆壳	平均35公里厚
洋壳	平均7公里厚
岩石圈	深达75公里
地幔(硅酸盐)	2900公里厚
	基部温度约3000°C
外核(液态铁)	2200公里厚
	基部温度约4000°C
内核(固态铁)	1200公里厚
	中心温度高达5000°C

着某种热力发动机,地幔中并非完全固态的岩石在数十亿年时间里缓慢循环。正是这种循环驱动了板块构造运动,导致大陆漂移,并引发了火山爆发和地震。

岩石循环

在地表,我们脚下的热力发动机与头顶的太阳灶彼此呼应,其协同作用驱动了岩石周而复始的循环。地幔循环和大陆碰撞所抬升的山脉受到了太阳能驱动的风、雨和雪的侵蚀。化学过程也在起作用。大气层的氧化,活生物所产生的酸的化学溶解,以及溶解的气体均有助于分解岩石。大量的二氧化碳可以溶解在雨水中,形成导致化学风化作用的弱酸,把硅酸盐矿物变成黏土。这些残余的岩石被冲回,沉在河口和海底,形成新的沉积物,最终被抬升形成新的山脉,或沉回地幔进入深层的再循环。整个过程是由结合进矿物结晶结构中的水来润滑的。这一岩石循环由18世纪的詹姆斯·赫顿[1]首先提出,但他当时并不清楚循环发生的深度及其时间尺度。

到目前为止,我们只是粗略介绍了我们这个神奇星球的皮毛而已。接下来我们就要启程,去挖掘岩石深处和那遥远的过去的秘密了。

---

[1] 詹姆斯·赫顿(1726—1797),英国地质学家、医生、博物学家、化学家和实验农场主。他在地质学和地质时期领域提出了火成论和均变论,为现代地质学的发展奠定了基础。

第二章
"深时"

> 太空很宽广。实在太宽广了……你或许觉得沿街一路走到药房已经很远了，但对于太空而言也就是粒花生米而已。
>
> ——道格拉斯·亚当斯，《银河系搭车客指南》

（图中文字按顺时针方向依次为：侏罗纪，白垩纪，第三纪）

世界不仅空间维度巨大，其时间之久远更是超出人们的想象。如果不了解约翰·麦克菲①、斯蒂芬·杰伊·古尔

---

① 约翰·麦克菲（1931— ），美国作家。他被视为创意纪实写作的先驱者，1999年获普利策奖。

德[1]和亨利·吉[2]等作家笔下的"深时"[3],就无法全面掌握地质学的各种概念和运作过程。

我们大都认识自己的父母,许多人还记得祖父母,但只有少数人见过曾祖父母。他们的青年时代比我们所处的时代早一个多世纪,那对我们来说相当陌生,因为我们的科学认识和社会结构都已大不相同。仅仅十几代之前,英格兰还在伊丽莎白一世的统治之下,机械化运输和电子通讯还是人们做梦也想不到的事,欧洲人才首次探索美洲。30个世代之前,距离现在就有1000年,那时诺曼人尚未入侵英伦。人类用连续的文字记载来追溯自己的直系祖先,也是那以后的事情了。我们或许能够利用考古学和遗传学知识大致分辨出当时我们的祖先是什么人、他们可能住在什么地方,但我们永远不可能确凿无疑。50个世代之前,罗马帝国正处于全盛时期。150个世代之前,古埃及大金字塔还未建成。大约300个世代之前的欧洲新石器时代,最后一个冰川期刚刚结束,基本的农业是当时最新的技术革命。考古学看来无法揭示我们的祖先当时身居何处,尽管通过比较母系遗传的线粒体DNA(脱氧核糖核酸)可以确定大致的区域。给这个数字再

---

[1] 斯蒂芬·杰伊·古尔德(1941—2002),美国古生物学家、演化生物学家、科学史学家与科普作家。他的主要科普作品有《自达尔文以来》《熊猫的拇指》等。

[2] 亨利·吉(1962—),英国古生物学家和演化生物学家,科学期刊《自然》高级编辑,著有《寻找深时:超越化石记录,进入全新的生命史》等。

[3] 地质时间概念,其所指的时间远远超过《圣经·创世记》中暗示的区区数千年,达到数十亿年。现代哲学意义上的"深时"概念由18世纪苏格兰地质学家、人称"现代地质学之父"的詹姆斯·赫顿提出。从那以后,现代科学经过漫长而复杂的发展,确定地球的年龄约为45.4亿年。

加一个0，我们就回到了3000个世代之前，也就是10万年前。在这个时期，我们无法追溯任何现存种族群体的独立血统。线粒体DNA表明，在那之前不久，所有的现代人类在非洲拥有一个共同的母系祖先。然而从地质时间的角度来看，此皆近世之事。

这个时间乘以10，也就是100万年前，关于现代人类的线索就无从查考了。再乘上10，就能看到早期类人猿祖先的化石遗迹。在如此久远的过去，我们无法指着某个单一的物种，肯定地说我们的祖先就在这些个体之间。再乘上10，即一亿年前，就是恐龙生活的年代。那时人类的祖先一定还是些类似鼩鼱的微小生物。10亿年前，也就回到了最初的化石之间，可以辨认的动物或许还一只都没有出现。100亿年前，太阳和太阳系还没有诞生，如今组成人类生存的星球和人类本身的原子，当时还在其他恒星的核反应堆中炙烤沉浮。时间的确深邃悠远。

再来考虑一下区区数代间可发生哪些变化。与地球的年纪相比，人类的历史微不足道，然而几个世纪就见证了多次火山喷发、惨烈地震和毁灭性的滑坡。再想想破坏性不那么剧烈的变化，它们一直绵延不断。在30个世代之内，喜马拉雅山脉的若干部分升高了一米或更多。但与此同时，它们受到的侵蚀很可能多于一米。一些岛屿诞生了，另一些则被淹没。一些海岸由于受到侵蚀而变矮了数百米，另一些却高耸于水面之上。大西洋加宽了大约30米。好了，把所有这些距今较近的变化乘以10、100或1000，就可以看到在地质学上的"深时"期间，宇宙间可能发生了什么。

洪水与均变性

人类从史前时期就注意到了化石遗迹。某些古代的石器经过打磨削尖的一番折腾，似乎单纯就是为炫耀那些贝壳化石。一个古代伊特鲁里亚①的墓穴中就放置了一株巨型苏铁类植物的树干化石。但了解化石性质的努力却从近代才刚刚开始。地质科学最初兴起于信仰基督教的欧洲，那时人们的信仰主要源自《圣经》故事，因而在山区高地发现已灭绝生物的贝壳和骨头并没有令他们吃惊：那些都是在《圣经》记载的大洪水中消失的动物的遗骸。所谓的水成论者甚至认为花岗岩是远古海洋的沉淀物。洪水之类本是上帝的极端行为，这一概念促使人们想象地球是由大灾难造就的，直至18世纪末，这一直是普遍接受的理论。

1795年，苏格兰地质学家詹姆斯·赫顿出版了《地球论》（*Theory of the Earth*），该书如今已成名著。书中常被引用的一句话（尽管是改述的概要）是："现在是通往过去的一把钥匙。"这是渐变论或均变论的理论，该理论认为，要想了解地质过程，就必须观察当前正在发生的那些几乎察觉不到的缓慢变化，然后只需在历史上加以追溯即可。查尔斯·莱尔②阐述了这一理论并始终为之辩护，他生于1797年，赫顿恰是在那年去世的。赫顿与莱尔两人都试图将对创世和洪水等事件的宗教信仰搁置一旁，提出作用于地球的渐进过程是无始无终的。

确定创世的年代

试图计算地球年龄的努力最初起源于神学。所谓的神创

---

① 公元前12世纪至前1世纪的意大利中西部古国，大致在如今的托斯卡纳地区。

② 查尔斯·莱尔（1797—1875），英国地质学家。他的理论多源自他前一代的地质学家詹姆斯·赫顿。他的巨著《地质学原理》影响巨大，查尔斯·达尔文在"小猎犬"号旅行期间就携带了此书。

论者照字面意义解释《圣经》,因而坚称神创造世界仅用了七个整天,要算是相对近代的事。圣奥古斯丁在其对于《圣经·创世记》的评注中指出,上帝的视野远在时间之外,因而《圣经》中所提到的创世期间的每一天都可能比24小时长得多。就连在17世纪被人们广泛引用、由爱尔兰的厄谢尔大主教做出的估计——地球是公元前4004年创造出来的——也只是为了估算地球的最小年龄,而且是基于对史料的仔细研究,尤其是好几代大主教和《圣经》中提到的诸位先知所记载的史料。

首次基于地质学估计地球年龄的认真尝试是1860年由约翰·菲利普斯所为。他估计了当前的沉积速率和所有已知地层的累积厚度,估算出地球的年龄将近9600万年。威廉·汤普森——后来的开尔文勋爵——继承了这一观点,基于地球从起初的熔融态炙热球体冷却所需的时间,做出了估计。值得一提的是,他起初估算出的地球年龄也是非常接近的数字,即9800万年,不过后来他进一步推敲,将其缩短至4000万年。但均变论者和查尔斯·达尔文认为他们估算的年代还是太近,基于达尔文提出的自然选择演化论,物种的起源需要更长的时间。

20世纪初,人们认识到额外的热量或许来自地球内部的放射现象。因此,基于开尔文的构想,地史学得以拓展。然而,最终促使我们如今对地球年龄进行日益精确估计的,还是对放射现象的理解。很多元素都以不同的形态或同位素的形式存在,其中一些具有放射性。每一种放射性同位素都有其独特的半衰期,在此期间,该种元素任意给定样品的同位素均可衰变一半。就这种同位素本身而言,这没有什么用处,除非我们知道衰变开始时的准确原子数量。但通过测量不同的同位素的衰变速率及其产物,就有可能得到异常精确的年代。20世纪早期,欧内斯特·卢瑟福①宣布,某种名为沥青铀矿的放射性矿物,其一份特定样品的地质年龄有7亿年之久,比当时很多人认为的地球年龄要长得多。此言一发即引起了巨大轰动。后来,剑桥大学物理学家R.J.斯特拉特②通过累计钍元素衰变所产生的氦气证明,一份来自锡兰(今斯里兰卡)的矿物样品的地质年龄已逾24亿年。

在放射性测定地质年代方面,铀是一种很有用的元素。铀在自然界有两种同位素——它们是同一元素的不同形式,差别仅在于中子的数量,因而原子量也有差别。铀238经由不同的中间产物最终衰变成铅206,其半衰期为45.1亿年,而铀235衰变成铅207,其寿命不过7.13亿年。对从岩石中提取的这四种同位素进行比率分析,加之以衰变过程中产生的氦气累计,就可以给出相当准确的年代。1913年,阿瑟·霍姆斯③使用这一方法,首次准确估计了过去6亿年间各个地质时期的持续年代。

放射性测定地质年代技术的成功在相当程度上得益于质谱仪的效力,这种仪器实际上可以将单个原子按重量排序,因而使用非常少量的样品即可给出痕量组分的同位素比率。但其准确性取决于有关半衰期的假设、同位素的初始丰富度,以及衰变产

---

① 欧内斯特·卢瑟福(1871—1937),出生于新西兰的英国物理学家,被誉为原子核物理学之父。

② 瑞利·约翰·斯特拉特(1842—1919),与威廉·拉姆齐合作发现了氩元素,并因此获得了1904年的诺贝尔物理学奖。

③ 阿瑟·霍姆斯(1890—1965),英国地质学家。他对于地质学的理解做出了两大贡献:其一是倡导放射性矿物确定年代的应用,其二是理解地幔对流的机械和热力学意义,这最终促成板块构造学说被广泛接受。

用于测定地质年龄的某些放射性同位素

同位素	产物	半衰期	用途
碳14	碳12	5730年	确定长达5万年前的有机残余物的年代
铀235	铅207	7.04亿年	确定侵入体和个体矿物颗粒的年代
铀238	铅206	44.69亿年	确定远古地壳中个体矿物颗粒的年代
钍232	铅208	140.1亿年	同上
钾40	氩40	119.3亿年	确定火山岩的年代
铷87	锶87	488亿年	确定坚硬的火成岩和变质岩的年代
钐147	钕143	1060亿年	确定玄武质岩和非常古老的陨石的年代

物随后可能发生的逸出。铀同位素的半衰期令其很适合用于测定地球上最古老的岩石。碳14的半衰期仅为5730年。在大气层中，碳14由于宇宙射线的作用而不断得到补给。一旦碳元素被植物吸收，植物死亡后，同位素不会再得到补给，从那一刻起，碳14的衰变就开始了。因此，用它来测量诸如考古遗址的树木年龄等再合适不过了。然而事实上，大气层中的碳14含量是随着宇宙射线的活动而变化的。正因为已知树木的年轮就可以独立计算出年代，我们才知道可以用碳14作为测定工具，并对长达2000年的碳定年予以校正。

地质柱状剖面

仔细观察某个崖面上的一段沉积岩，能看出它包含若干层。有时，与洪水和干旱相对应的年层是肉眼可见的。更多时候，地层代表成千上万甚至数亿年间偶然发生的灾难性事件，或者缓慢而稳定的沉淀，紧随其后发生的环境变化会导致岩石层略有不同。如果古岩石片段纵深很长，像美国亚利桑那州的大峡谷那样，则表示有数亿年的沉积。人类天生喜爱分门别类，多层沉积岩显然很能迎合这一癖好。但在观察一个体量狭窄的平层崖面时，人们很容易忘记这些岩层在全世界范围内并不是连续的。整个地球从来没有被类似沉积岩那样的单一海洋浅覆层覆盖过！正如现今地球上有河流、湖泊，还有海洋、沙漠、森林和草原，远古时期也一样，那时地球上也存在着一系列壮观的沉积环境。

地质年代的主要分期

19世纪初，英国土木工程师威廉·史密斯首先了解到这一点。他当时在为英国的新运河网勘探地形，发现国内各地的岩石有时会包含相似的化石。在某些情况下，岩石的类型相同，而有时只是化石相似。史密斯以此为依据，为不同地方的岩石建立关联，并设计出一个全面的序列。最终，他发表了世界上第一张地质图。20世纪人们又测定出很多地质年代，加之不同大陆间的岩石被关联起来，就能够发布一个单一岩层序列，用来代表整个世界范围内的各个地质时期了。我们如今所知的地质柱状剖面是多种技能相结合的产物，推敲经年，并经国际协作，达成了一致意见。

地质年代的主要分期

宙	代	纪		世	
显生宙	新生代	第四纪		全新世	0.01
				更新世	1.8
		第三纪	晚第三纪	上新世	5.3
				中新世	23.8
			早第三纪	渐新世	33.7
				始新世	54.8
				古新世	65.0
	中生代	白垩纪		晚白垩世	
				早白垩世	142
		侏罗纪		晚侏罗世	
				中侏罗世	
				早侏罗世	205.7
		三叠纪		晚三叠世	
				中三叠世	
				早三叠世	248.2
	古生代	二叠纪		晚二叠世	
				早二叠世	290
		石炭纪	宾夕法尼亚亚纪	晚石炭世	323
			密西西比亚纪	早石炭世	354
		泥盆纪		晚泥盆世	
				中泥盆世	
				早泥盆世	417
		志留纪		晚志留世	
				早志留世	443
		奥陶纪		晚奥陶世	
				中奥陶世	
				早奥陶世	495
		寒武纪		晚寒武世	
				中寒武世	
				早寒武世	545
前寒武纪	远古宙				2500
	太古宙				4000
	冥古宙				4560

图5 地质年代的主要分期(不按比例)。所列年代(位于右侧,以距今百万年为单位)是2000年国际地层委员会表决通过的

灭绝、非均变，以及大灾难

显然，地质柱状剖面中的某些变化更加剧烈，人们根据这一便利，将地质历史划分为不同的代、纪和世。有时，岩石性质会发生突然而显著的变化，跨越了某一地质历史界限，这表明环境出现了重大变化。有时会发生所谓的非均变，是由诸如海平面变化等原因导致的沉积作用中断，因而要么沉积作用终止，要么岩层在柱状剖面延续之前便被侵蚀殆尽。化石所记录的动物区系的重大变化也是这类突变的标志，很多物种灭绝了，新的物种开始出现。

地质记录中出现的几次间隔突出显示了其间发生的严重的大规模物种灭绝。寒武纪末期和二叠纪末期都以海洋无脊椎动物中将近50%的科类物种和高达95%的个体物种的灭绝为标志。在标记了三叠纪后期和泥盆纪后期的物种灭绝期间，分别有大约30%和略低于26%的科类物种消失了，但是，距今最近也最著名的大规模消亡则发生在6500万年前的白垩纪末期。所谓的K/T界线[①]之所以举世皆知，不仅因为在此期间最后一批恐龙灭绝，还因为它为该物种灭绝的原因提供了充分证据。

来自太空的威胁

沃尔特[②]和路易斯·阿尔瓦雷茨[③]首先提出，恐龙灭绝可能

---

① 白垩纪与第三纪的界线。
② 沃尔特·阿尔瓦雷茨（1940— ），美国地质学家，任教于加州大学伯克利分校地球与行星科学系，因与其父共同提出恐龙灭绝是因为小行星或彗星撞击地球的理论而闻名。
③ 路易斯·阿尔瓦雷茨（1911—1988），沃尔特·阿尔瓦雷茨的父亲，西班牙裔美国实验物理学家，1968年获诺贝尔物理学奖。

是天体碰撞的结果，此提法起初并没有多少科学依据。但是，他们随即发现，在地质柱状剖面中的那一个时间点上，沉积物窄带中富含铱，这是某些类型的陨石中富含的元素。但没有发现那一时期的撞击坑的迹象。再后来，证据开始出现，不是来自陆地，而是在墨西哥尤卡坦半岛离岸很近的海中，这一掩在海下的撞击坑直径达200公里。在广阔得多的区域还发现了碎片的证据。如果像科学家们计算的那样，这一地点标记了直径或达16公里的小行星或彗星撞击地球的位置，其结果的确会是毁灭性的。除了撞击本身的影响及其所导致的海啸，如此众多的岩石也将会蒸发并散布在地球的大气层中。起初，气候会无比炎热，辐射热会引发地面上的森林火灾。灰尘会在大气层中停留数年之久，遮天蔽日，造成环球严冬，导致食用植物和浮游生物大量死亡。撞击地点的海床含有富硫酸盐矿物的岩石，这些物质会蒸发，并导致致命的酸雨，从大气层中冲刷而下。如果发生了这样的灾难还有任何生物幸存下来，才真是令人称奇。

来自内部的威胁

我们曾一度很难理解物种大规模灭绝究竟是如何发生的，而如今出现了很多彼此对立的理论，又让人不知道该信哪个好了。这些理论多涉及剧烈的气候变化，有些由宇宙撞击或海平面、洋流和温室气体的变化所引发，还有些是由诸如漂移或重大的火山活动等地球内部的原因所导致的。的确，我们所知的大多数物种大灭绝似乎至少都与溢流玄武岩大喷发大致同时发生。在白垩纪末期，正是这些大喷发在印度西部产生了德干地盾。甚至还有一种观点认为，小行星的一次大撞击引起了在地球另一侧聚焦的冲击波，从而引起了大喷发。但是时间和方位看来并不足以支持那种解释。无论是何原因,生命和地球的历史总是不时被一些灾难性事件打断。

混乱占了上风

我们都记得过去十年左右所经历的极端气候事件,像是最寒冷的冬天、洪水、暴风雨,或是干旱等等。若把时间推回到一个世纪之前,恐怕只有那些更大的事件才会令人印象深刻。专家在规划水灾的海岸维护与河流防线时,经常会使用"百年不遇"的概念;这些防线的设计要经得起百年一遇的洪水才行,它们很可能比十年一遇的洪水要严重得多。但如果把考察的时间延展到1000年或100万年,总还会有更大、更严重的事件。据某些理论家所言,从水灾、暴风雨和干旱到地震、火山喷发和小行星撞击皆是如此。在地质历史的尺度上,我们可要小心点儿才是!

更加深邃的时间

书本中经常列出的地质时期表只会回溯到大约6亿年前寒武纪开始的时间,而忽略了我们这个星球40亿年的历史。正如美国加州大学的比尔·舍普夫[①]教授所说,大多数前寒武纪岩石的问题,在于它们无法鉴定——乱七八糟,没有任何方法可以识别。地球内部持续的构造再处理,以及地表风化和侵蚀的不断打击,意味着好不容易幸存下来的大多数前寒武纪岩石都有着严重的折痕和变形。不过,在大多数晴朗之夜,人们都可以看见有40多亿年历史的岩石——得要举头望明月,而不是低头看地球。月球是个冰冷死寂的世界,没有火山和地震、水或气候来改

---

① 比尔·舍普夫(1941—),美国古生物学家,在加州大学洛杉矶分校教授地球科学。

换新颜。它的表面覆盖着陨石坑，但其中大多数都是在月球形成的早期发生的，当时太阳系里还充满着飞散的碎片。

　　至于在地球上幸存下来的前寒武纪岩石，它们诉说着一个古老而迷人的故事。它们并不像达尔文猜想的那样全无生命的痕迹。的确，在前寒武纪末期，从大约6.5亿年前到5.44亿年前，曾经出现了各种怪异的化石，特别是在澳大利亚南部、纳米比亚和俄罗斯等地。在那以前似乎有过一个特别严酷的冰河作用时期。有人使用了"雪球地球"这种说法，表示环球的海洋在当时有可能全部冻结。对生命而言，那必定是一个重大的挫折，并且没有多少证据能够证明在此之前出现过多细胞生命形式。但大量证据表明那时已经出现了微生物——细菌、蓝藻细菌和丝状藻类。澳大利亚和南非有距今大约35亿年的丝状微化石，而格陵兰岛38亿年高龄岩石中的碳同位素看上去也像生命的化学印记。

　　在起初的7亿年历史中，地球一定特别荒凉。当时有为数众多的大撞击，剧烈程度远甚于或许造成了恐龙灭绝的那一次。后一次重轰炸期的疤痕还能在月球上的月海中看到，那些月海本身就是巨大的陨石坑，充满了撞击所熔融的玄武岩熔岩。这样的撞击会熔融大量的地球表面，并无疑会把任何原始海洋蒸发殆尽。如今我们星球上的水很可能来自随后的彗星雨及火山气体。

生命的曙光

　　人们一度认为，地球早期的大气层是甲烷、氨、水和氢的气体混合物，这是组成原始生命形式的碳的潜在来源。但如今人们认为，来自年轻太阳的强烈紫外线辐射迅速分解了那种气体混合物，产生了二氧化碳和氮气的大气层。生命的起源仍是个未解之谜。甚至有人认为，生命可能源自外星，是来自火星或者更远星球的陨石抵达地球后带来的。但当前的实验室研究表明，某些化学体系可以开始进行自我组织并催化其自身的复制。有关现今生命形式的分析指出，最原始的生命并非那种以有机碳为食，或利用阳光助其光合作用的细菌，而是如今在深海热液喷溢口发现的那种利用化学能的细菌。

到35亿年前，几乎必然存在着微小的蓝藻细菌，多半也已经出现了原始藻类——就是我们如今在死水塘中看到的那种东西。这些生物开始产生了戏剧性的效果。它们利用阳光作为光合作用的动力，从大气层中吸收二氧化碳，有效地侵蚀着二氧化碳保护层，这可是在太阳作用较弱时通过温室效应给地球保暖的。这或许最终导致了前寒武纪末期的冰川作用。但在此很久以前，这些生物的作用就已经导致空前绝后的最糟糕的污染事件。光合作用释放了一种此前从未在地球上存在过的气体——氧气，对很多生命形式可能是有害的。起初，氧气不能在大气层中长期存在，但它很快就与海水中溶解的铁元素发生反应，产生了条带状氧化铁的密集覆层。整个世界都生锈了——这可不是什么夸张的说法。但光合作用仍在继续，大约24亿年前，游离氧开始在大气层中逐渐积累，为可以呼吸氧气和进食植物的动物生命的到来铺平了道路。

地球的诞生

大约45亿年前，曾有一大片气体尘埃云，这是以前若干代恒星的产物。在重力的影响下，这片云开始收缩，其过程或许还由于附近某颗恒星爆炸或超新星的冲击波而加速。随着这片云的收缩，其内的轻微旋转加速，将尘埃散布出去，在原始星体周围形成扁平的圆盘。最终，主要由氢和氦组成的中心物质收缩到足以在其核心引发核聚变反应，太阳开始发光。一阵带电粒子的风开始向外吹，清除了周围的部分尘埃。在这片星云或圆盘的内部，只剩下耐火的硅酸盐。在远处，氢和氦加速形成了庞大的气体行星：土星和木星。水、甲烷和氮等挥发性冻结物被推到更远处，形成了外行星、柯伊伯带①天体和彗星。

内行星——水星、金星、地球和火星——是由已知的增积过程形成的，起初粒子彼此碰撞，有时会裂开，偶尔也会彼此联合。最终，较大的粒子团积累足够的重力引力，把其他粒子团拉向它们。随着体积的增长，撞击的能量也增加了，撞击熔融了岩石并导致其成分析出，其中密度最大、富含铁元素的矿物质下沉形成了核心。在撞击、重力收缩释放的能量，以及放射性同位素衰变等多重作用之下，崭新的地球变得十分炙热，大概至少有一部分被熔解了。前太阳星云②中的很多放射性元素可能在超新星爆炸前不久就产生了，仍因其具有放射性而十分炙热。因此，地球表面起初很难有液态水存在，而且最初的大气层可能多半都被太阳风吹散了。

---

① 太阳系海王星轨道外黄道面附近天体密集的中空圆盘状区域，以荷裔美籍天文学家杰拉德·柯伊伯（1905—1973）命名。

② 科学家通过对古陨石的研究，发现了短暂同位素（如铁60）的踪迹，该元素只能在爆炸及寿命较短的恒星中形成。这表示在太阳形成的过程中，附近发生了若干次超新星爆发。其中一颗超新星的冲击波可能在分子云中造成了超密度区域，导致该区域塌陷。这种塌陷气体区域被称为"前太阳星云"，其中的一部分形成了太阳系。

青出于蓝

长期以来,月球的形成对科学界来说一直是一个谜。人们一度认为月球是从年轻的地球中分离出去,在地球旁边形成,或在经过地球时被其捕获的,但月球的组成、轨道和自转与这种说法并不吻合。然而现在有一个理论很合乎情理,也用计算机模型进行了很有说服力的模拟。该理论指出,一个火星大小的原行星曾在太阳系形成大约5000万年之后与地球发生了碰撞。这一抛射物的核心与地球的核心相融合,撞击力熔融了地球的大部分内部物质。撞击物的大部分外层,连同某些地球物质一起被蒸发并投入太空。其中的许多物质聚集在轨道中,累积合生,形成了月球。这次灾难性事件让我们收获了一个良朋挚友,它对于地球似乎有着稳定的作用,防止地球自转轴的无序摇摆,因而让我们的行星成为生命更宜居的家园。

第三章
地球深处

地球的表面覆盖着一层相对较薄的冷硬外壳。在海洋下面，这层外壳大约有七八公里厚，而就大陆而言，其厚度则是30—60公里。其基部是莫氏不连续面[①]，又称"莫霍面"，它可以反射地震波，这大概是由于它的组成发生了变化，变成了其下地幔的致密岩石。岩石圈是地球表面上一层冰冷坚硬的物质所组成的完整板块，不但包括地壳，还包括地幔的顶部。大陆岩石

---

① 1909年由克罗地亚地震学家安德里亚·莫霍洛维契奇（1859—1936）首先发现。

圈总共有大约250公里乃至300公里厚。海洋下的岩石圈较薄，越接近洋中脊越薄，最薄处仅比7公里洋壳略厚。然而岩石圈并不是一个单一的坚硬地层，它可以分成一系列所谓的构造板块。这些板块是我们了解地球深处如何运作的主要线索。为了解那里发生的事，我们必须深入地壳之下一探究竟。

深度挖掘

在距离我们只有30公里的地方，有一个我们永远不能探访的所在。30公里的横向距离不过是一次轻松的公交之旅，但在我们脚下，这一距离几乎就是一个难以想象的高温高压之处了。任何矿井都不可能开采到如此之深。1960年代，有人提议利用石油开采业的海洋钻探技术，直接钻通洋壳进入地幔，这就是所谓的"莫霍计划"，后来由于成本之巨和任务之艰而未能实施。在俄罗斯的科拉半岛和德国境内进行的深层钻探尝试在达到大约11 000米深度后就放弃了。这不仅是因为岩石难以钻孔，而且热量和压力会软化钻机的部件，还会把刚刚钻开的孔洞立即重新压合。

来自地球深处的信使

我们可以通过一种方法直接从地幔中取样：利用深源火山的喷发物。火山喷发出来的岩浆大多只是来自源地物质的部分熔融物，因此，举例来说，玄武岩并非表层岩的完备样本。然而，它却能够提供关于其下物质的同位素线索。例如，某些来自夏威夷等地的深源火山的玄武岩中含有氦3及氦4比率较高的氦气，据信早期太阳系的情况也是如此。因此人们认为，这种玄武岩来自地球内部的某个至今仍然保持着本来面貌的部分。火山喷发时氦逸失了,被放射性衰变所产生的氦4缓慢地取代。洋脊火山玄武岩中的氦3耗尽了。这意味着这种玄武岩是再生物质,其氦气在早期的喷发中逸失,而且这种玄武岩并非来自地幔深处。

剧烈的火山喷发有时的确会在其岩浆中携带着更直接的表层岩样本。这些所谓的"捕虏岩"是熔岩流携带而出的尚未熔融的表层岩样本。它们通常是诸如橄榄岩等暗绿色的致密岩石,富含橄榄石矿,后者是一种镁铁硅酸盐。山脉深处有时也会找到类似的岩石,是从地球极深处强推出地面的。

慢速熔岩流

坎特伯雷座堂①那富丽堂皇的中世纪彩色玻璃窗可以透露一些有关地球地幔性质的信息。窗子由很多小块的彩色玻璃组成,嵌在跨距很大的窗框里。如果观察透过窗格玻璃的阳光,就会注意到底部的光线比顶部的要暗一些。这是由于玻璃的流动。用专业术语来说,玻璃是一种过冷的流体。历经若干世纪,重力令窗格缓慢下垂,底部的玻璃因而会较厚一些。然而,如果用手摸或者锤击(求主饶恕!),玻璃仍然呈现出固体样态。了解地球地幔的关键在于认识到,那里的硅酸盐岩石能够以同样的方式流动,尽管它们并未熔融。实际上,这些个体矿物颗粒一直在重新形成,从而引起了被称作"蠕动"的运动。结果是地幔极具黏性,就像非常黏厚的糖蜜。

---

① 位于英国东南部肯特郡坎特伯雷市,是英国圣公会首席主教坎特伯雷大主教的主教座堂,也是英国最古老、最著名的基督教建筑之一。

地球的全身扫描

关于地球的内部结构，最明确的线索来自地震学。地震通过星体发出地震冲击波。就像光线被透镜折射或被镜子反射那样，地震波穿行于地球，并在其不同的地层中反射。随着岩石的温度或软度不同，地震波的行进速度也有差异。岩石温度越高就越软，冲击波行进的速度也就越慢。地震波主要有两种，初至波（P波）速度更快，因而会率先抵达测震仪；另一种是续至波（S波）。P波是进行推拉运动的压力波；S波是剪切波，无法在液体中行进。正是通过对S波的研究，人类首次揭示了地球的熔融外核。在单一的仪器上探测这些地震波并不会显示多少信息，但如今全球各地散布着由数以千计灵敏的测震仪组成的网络。每天都有很多小型地震发出信号。结果有点像医院里的全身扫描仪，患者被X射线源和传感器环绕，计算机利用结果来构建患者内部器官的三维影像。医院的探测装置被称作计算机辅助断层扫描，而地球的相应版本则被称作地震层析成像。

测震仪的全球网络最适于在全球范围内观测事物。它会揭示地幔的整体分层，以及每隔数百公里，地震波速因温度高低而发生的变化。世界上还部署着间隔更小的矩阵，起初设置它们的目的是探测地下核子试爆，它们连同地球物理学家部署在感兴趣的地质区域的新型矩阵一起，有望观测到数公里之深的地幔结构。并且，似乎每一个尺度上都有自己的构造。在这些地球全身扫描中，最清楚的莫过于地层了。在2890公里——我们这个星球液态外核的厚度——之下S波无法通过。但是地幔有几个显著的特征。比如前文中提到的，地壳基部有莫氏不连续面，另一个则位于坚硬的岩石层的基部。岩石层下的软流层比较软，因而地震波速较慢。410公里以下和660公里以下分别是界限清晰的地层，而520公里深处左右则是一个不甚清晰的地层。在地幔基部还有另一个被称作核幔边界（D″分界层）的地层，它很可能是不连续的，厚度范围从0公里到大约250公里不等。

地震层析成像同样揭示了一些更加微妙的特征。本质上，较为冰冷的岩石也会更硬，因而相对于较热较软的岩石，地震波在其间行进的速度也更快。在古老而冰冷的洋壳插到大陆之下或者插进海沟的位置，下沉板块的反射显示出其通向下方地幔的通道。在那里，地球炙热的核心烘烤着地幔的底面，将其软化并抬升成一个巨大的地柱。

地幔充满了未解之谜，它们乍看上去像是彼此矛盾的。它是固态的，却可以流动。它由硅酸盐岩石组成，硅酸盐岩石本是一种优良的隔热体，但不知为何，却有大约44太瓦特[①]的热量穿过地幔涌向地表。很难弄清热流是如何仅靠传导完成的，然而如果确实存在对流，地幔就应该是混合物，那么它又如何能显示出分层构造？此外，除非地幔中存在未混合的区域或地层，海洋火山喷发出的岩浆中所含的示踪同位素混合物，为何全然不同于据信存在于地幔主体内的混合物？这些谜题是近年来地球物理学的主要研究领域之一。

地幔上的钻石窗口

某些最有用的线索来自对地下那些岩石性质的了解。为了查明地球深处那些岩石的状况，就必须复制那里令人难以置信的巨大压力。令人惊异的是，这种情形动动手指就能模拟：握住

---

① 功率单位，1太瓦特=10^{12}瓦特。

两颗优质的宝石级金刚石，用珠宝商的术语来说，其切工为"明亮型"，即每颗金刚石的顶部均有一个完全平坦的小切面。将一个微小的岩石样本置于两个小切面中间，然后用指旋小螺钉将两个小切面拧得更紧一些。两个金刚石砧之间的力非常集中，以至于仅仅拧动螺钉，产生的压力就会超过300万个大气压（300吉帕斯卡[①]）。金刚石是透明的，通过激光照射就可以对样本加热，也方便使用显微镜和其他设备来观测。这实际上可以作为一个窗口，让我们观察地幔深处的岩石的状况。

  一天，比尔·巴西特教授在康奈尔大学的实验室里研究金刚石砧上的一个微小晶体。当他提高压力时，没有发生什么变化，于是他决定先去吃午餐。正要离开，忽听到砧上传出突如其来的爆裂声。显然，他的宝贝金刚石碎了一颗，他冲回去在显微镜下仔细观察。钻石安然无恙，但样本突然变成了一种全新的高压晶体形式。这就是所谓的相变：组分依然，但结构发生了变化——在该例中，变成了更加致密的晶格。

  从捕虏岩的组分我们知道，至少上地幔是由橄榄岩等岩石组成的，其中富含镁铁硅酸盐矿物橄榄石。把这种岩石的微小样本放在金刚石砧中间加压，它就会经历完整的一系列相变。在相当于410公里地幔深处所受的大约14吉帕斯卡的压力之下，橄榄石变形成一种名为瓦兹利石的新结构。在相当于地下520公里、18吉帕斯卡的压力下，它又发生了形变，变成了林伍德石，一种尖晶矿石的形式。然后，在23吉帕斯卡，相当于地下660公里所受的压力之下，会变成两种矿物：一种是钙钛矿，另一种是名为镁方铁矿的镁铁氧化物矿物。我们会注意到，相变发生的深

[①] 压强单位，1吉帕斯卡=10^9帕斯卡。

度恰恰是可以反射地震波的地方。因此，这些地层或许可以表示晶体结构的变化，而非组分的变化。

双层蒸锅？

地下660公里的地层是上地幔和下地幔的分界线，这是一个特别强烈的特征，也是学界激烈辩论的焦点：有人认为整个地幔都是在一个巨大的对流系统中循环，而另一些人认为地幔更像是一口双层蒸锅，上、下地幔各有其独立的循环腔，两个腔体之间几乎没有物质交换。历史上，地球化学家更偏爱双层结构，因为这种结构考虑到了不同地层之间的化学差异，而地球物理学家则偏爱全地幔对流。当前的迹象表明，两者可能都是正确的，在这一折中方案中，全地幔循环是可能的，但绝非易事。地震层析成像的数据初看之下或许偏向双层蒸锅的观点。地震扫描揭示了下压的洋壳板块沉向660公里处异常地层的位置，但它们似乎并未通过该地层。相反，物质散布开去，似乎又在数亿年间在该深度聚集起来。但进一步扫描显示，在某一位点，下压的洋壳板块可以像雪崩一样突破并继续沉到下地幔，直至地核的顶部。

1994年6月，玻利维亚遭遇了一场剧烈的地震。地震几乎没有造成破坏，因为震源很深——大约有640公里。但在那样的深度，岩石应该是太软了，以至于无法断裂。正是在地震发生的区域，来自太平洋古老洋壳的一个板块下沉到安第斯山脉以下。当时，想必是一整层的岩石经历了一场灾难性相变，变成了更加致密的钙钛矿结构，它似乎必须经历这样的变化，才能够沉入下地幔。这个解释一举解开了地幔分层和深层地震的秘密。

但仍有很多问题有待解释。例如，潜入太平洋汤加海沟之下

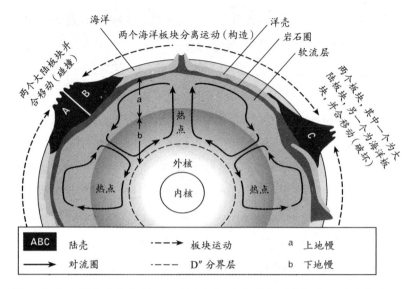

图6 地球地幔内的基本循环，及其如何反映在岩石圈板块运动和各个板块边界上。为清楚起见，运动均已简化，岩石圈的纵向比例尺也放大了不少

的洋壳板块以每年250毫米的速度穿过地幔，对于其温度而言，这一速度过快，无法稳定下来。洋壳物质会在区区300万年内抵达上地幔的基部，如果淤积在那里或者延展至下地幔，其低温问题理应十分突出。但没有证据表明存在着这样的板块。某个理论声称，不是所有的橄榄石都转变成了密度更高的矿物，因而原本的板块会中立地漂浮在上地幔中。低温加上矿物成分使得其地震波速与其他地幔物质非常相近，因此，它不会轻易显现出来，就像一层甘油不会在水中突出显露一样。事实上的确存在着诱人但微弱的地震学证据，证明在斐济的下面存在着这样的板块。

钻石中的讯息

钻石是高压形式的碳,只会形成于地球100多公里深度之下,有时还要比这深得多。钻石中的同位素比率表明,下潜的洋壳中经常会有碳形成的钻石,或许是来自海洋沉积物中的碳酸盐。钻石中有时会含有其他矿物的微小内含物。宝石商人多半不怎么欢迎这个特征,但地球化学家对此求之不得。对那些内含物的精密分析,可以揭示有关钻石形成和穿过地幔的漫长的、有时颇迂回曲折的历史。

有些钻石内含一种名为顽火辉石的矿物,这是硅酸镁的一种形式。一些研究人员认为,它原本是来自下地幔的硅酸镁高温钙钛矿。证据是,根据他们的观察,这种矿物中所含的镍只有上地幔应有含量的十分之一。在下地幔的温度和压力下,镍被一种名叫铁方镁石的矿物所吸收,从而使得硅酸镁高温钙钛矿几乎不含镍——铁方镁石也是一种常见的钻石内含物。在少数情况下,内含物富含铝,在上地幔的环境里,铝被锁在石榴石之中。还有些内含物是富铁的,由此可知它们可能产生于地幔中靠近地核分界线的极深处。这些深处的钻石同样拥有一个与众不同的碳同位素特征,据信这是深层地幔岩而非潜没海洋岩石圈的特点。对钻石及其周围岩石的地质年龄的估计表明,其中的一些穿过地幔的道路漫长而迂回,或许用了逾十亿年时间。但这是个颇有说服力的证据,说明在上、下地幔之间至少存在着一些物质传递。

钻石被发现时所依附的岩石可不比钻石本身逊色多少。这种岩石名为金伯利岩,是以南非的钻石矿金伯利镇命名的。岩石本身简直是一团糟!除了钻石之外,它还含有各种各样不同岩石的多角团块和粉碎片段；这所谓的角砾岩是一种火山岩，往往会在远古的火山口形成胡萝卜形的岩颈。其确切组成很难判断，因为它在穿过岩石圈时吸收了太多的粉碎岩屑，但其原始岩浆一定主要由来自地幔的橄榄石和现在以云母形式存在的大量挥发性物质共同组成，其中橄榄石占大部分。如果它从地幔缓缓上升，如今我们就没有钻石了。钻石在地下不到100公里处的压力之下并不稳定，假以时日，它会在岩浆中熔解。但金伯利岩火山无暇等待。人们估计，物质穿过岩石圈的平均速度大约为70公里每小时。火山口在靠近表面的位置岩颈加宽，这表明挥发性物质正在剧烈扩张，表面的喷发速度可能是超音速的。因此，一路向上所采集的所有岩石碎片都猝熄了，它们凝固在时间中，因而成为来自岩石圈乃至地幔深处的各种岩石的样本。

地幔基部

近期对全球地震数据的分析显示，地幔基部有一个厚度最多200公里的薄层，即D″分界层。它不是一个连续的地层，而更像是一系列板块，也有点像地幔底面的一块块大陆。这里可能是地幔中的硅酸盐岩石与来自地核的富铁物质部分混合的区域。但另一种解释认为，这里是远古海洋岩石圈的长眠之所。在其沉降穿过地幔以后，板块依然冰冷致密，因而散布在地幔的基底，被地核缓慢加热，直到或许十亿年后，它以地幔柱的形式再次上升，形成新的洋壳。

根据测量，昼长存在微小差异，这同样提供了关于地球腹地的线索。因为月球对潮汐的牵引，以及最后一次冰河期的冰体压迫导致陆地上升，我们这个旋转的星球正在逐渐减慢转速。但仍有十亿分之一秒量级的微小差异。其中的一些或许应归因于大气循环吹到山脉上,就像海风吹鼓船帆。但另一部分看来像洋流推动船只的龙骨那样,是在外核中推动地幔基部脊线的循环引起的。因此,地幔的基部可能存在着像倒置的山脉那样的山脊与河谷。菲律宾地下十公里的地核中似乎有一个大洼地,其深度是美国大峡谷的两倍。阿拉斯加湾地下的鼓起是地核上的高点,那是个比珠穆朗玛峰还高的液态山峰。下沉的冰冷物质或许会在地核上压出凹痕,而热点则会向上鼓起。

超级地柱

下地幔的钙钛矿岩石尽管要炙热得多,却远比上地幔风化岩更有黏性。据估计,它的抗流动性要高出30倍。因此,地幔基部的物质以缓慢得多的速度上升,上升形成的柱体也要比上地幔的典型地柱粗大得多。它的表现很像熔岩灯里的黏性浆团,流动得极为缓慢。尽管某些物质在整个地幔中循环,但很可能真的存在一些只有上地幔才有的小对流圈。实验系统中对流圈的宽度趋向于与其深度一致,而至少在世界上的某些地区,地幔物质所组成的柱体,其间隔似乎与上地幔660公里的深度一致。

地球如何熔融

物质升降起落,生生不息。炙热的地幔岩地柱缓慢升向地壳,所受的压力也随之减轻,它们开始熔融。科学家们可以利用巨大的水压来挤压在熔炉内加热的人造石,再现当时的场景。岩石并非整体熔融,而只有一小部分如此;所产生的岩浆不像地幔其他部分那样稠密,因而得以快速上升到表面,作为玄武岩熔岩而被喷发出来。至于它是如何流经其他岩石的,曾经是另一个大谜团,最终的答案与岩石的精微结构有关。如果在岩石颗粒间形成的小熔融袋顶角很大，岩石就会像一块瑞士干酪；熔融袋不会相互连通，熔融物也不会流出来。但那些顶角很小，这样岩石就像一块海绵，所有的熔融袋也是相互连通的。挤压海绵，液体就会流出来。挤压地幔，岩浆就会喷发。

自由落体

艾萨克·牛顿看见苹果落下来，意识到重力会把物体拉向地心。但他不知道，在世界的某些地方，苹果下落的速度会比其他地方略快一些——人们通常不会注意到这一差异，也无法用苹果轻易测量出来。但宇宙飞船可以做到这个。根据道格拉斯·亚当斯在《银河系搭车客指南》中的说法，飞行的秘密，就在于一直在下落，却忘记了着陆。卫星差不多就是这样。卫星自由落下，但它的速度将其保持在轨道中。致密岩石区域更加强劲的万有引力会使卫星加速。在通过重力较小的区域时，卫星则会降速。通过跟踪低空卫星的轨道，地质学家可以绘制出位于轨道下方的地球的重力图。

在比较地球表面的重力图与地球内部的地震层析成像扫描图之后，结果大大出乎地球物理学家的意料。人们本来期待看到，冰冷致密的洋壳板块或因其密度更大而导致过大的地球引力，而由炙热地幔岩所组成的向上升起的地柱密度较小，因而其重力也较小。现实与此相反。情况在南部非洲尤为显著，那里似乎有个巨大的炙热地幔柱在上升，而在印度尼西亚附近，冰冷的板块正在下沉。麻省理工学院的布拉德·黑格对此做出了自己的解释。南部非洲地下的超级地幔柱正在导致相当一大块陆地上升，其上升的高度超出了预期——人们原以为大陆只是在静态的地幔上漂浮。他估计，跟自然漂浮在地幔上的位置相比，南部非洲被抬升了大约1000米，岩石的这种过度抬升导致重力增大。与此相似，印度尼西亚地下潜没的海洋岩石圈把周围的地表均拽在身后，造成重力较小，并导致海平面整个比陆地高出一块。如今就职于亚利桑那大学的克莱门特·蔡斯发现还有各种其他重力异常现象对应了曾经发生过的潜没事件。从加拿大的哈德孙湾，经由北极，穿过西伯利亚和印度，直到南极圈有一个很长的低重力带，似乎可以标记一系列潜没带，在过去的1.25亿年，那里的远古海床插入地幔。人们曾经认为是海平面上升导致了澳大利亚东半部的大部分大陆在大约9000万年前被淹，事实上或许是因为该大陆漂浮在一个远古潜没带上，在其经过时被潜没带拖曳，从而把陆地降低了600多米。

地核

我们不可能对地球的核心有直接体验，也不可能获得地核的样本。但我们的确从地震波中获悉，地核的外部是液态的，只有内核才是固态的。我们还知道，地核的密度远高于地幔。太阳系里唯一既有足够的密度，又储量丰富，足以组成地核那么大体积的物质就是铁。虽然没有地球核心的样本，我们却能在铁陨石中找到可能与之相似的东西。铁陨石不像石质陨石那样常见，但它更易于辨认。它们据信源自体积较大的小行星，早在这些小行星被发生在太阳系早期的轰炸粉碎之前，其铁质核心便分离了出来。它们的成分大部分是金属铁，但也含有7%至15%的镍。它们往往具有两种合金的共生晶体结构，一种含有5%的镍，另一种含有40%的镍，按比例组成了星球核心的总成分。

在新生的地球至少还是半熔融体时，铁质的地核必定已经通过重力从硅酸盐地幔中分离出来而形成了。随着地层的分离，诸如镍、硫、钨、铂和金等能够在铁水中熔融的所谓亲铁元素会与地层分离。亲岩元素则与硅酸盐地幔一起保留下来。铀和铅等放射性元素是亲岩的，而它们的衰变产物，或称子体，是铅和钨的同位素，因而会在地核形成之时被分离出来，进入地核。这必然会在地核形成时重置地幔中的放射性时钟。对于地幔岩地质年代的估计把这一分离的时间确定在45亿年前，大约比最古老陨石的地质年代晚5000万至1亿年，而最古老的陨石似乎出现在太阳系整体形成之时。

内核

地球的中心是冰冻状态，至少从铁水的角度来看，在地下难以置信的压力下，它是冰冻状态。随着地球的冷却，固态铁从熔融态地核中结晶出来。当前学界的理解是，产生地球磁场的发电机需要一个固态铁核，但纵观地球历史，它未必自始至终都有一个铁核。至于地球过去的磁场，其证据深锁于显生宙各个时期的岩石中。但大多数前寒武纪的岩石已经大变样了，因而很难测出其原本的磁性。这样一来，估计内核地质年龄的唯一方法，就只能来自地球缓慢冷却过程中地核热演化的模型。其计算方法与开尔文勋爵在19世纪末以地球冷却的速率来估算其年龄的方法相同。但现在我们知道，放射性衰变还会产生额外的热量。最新的分析表明，内核大概是在25亿至10亿年前这段时间开始固化的，取决于其放射性物质的含量。听上去或许是一段漫长的时间，但这意味着地球在其早期的数十亿年里是没有内核的，或许也没有磁场。

如今，内核的直径大约是2440公里，比月球的直径小1000公里。但它仍在持续增大。铁元素以大约800吨每秒的速率结晶，会释放相当数量的潜热，这部分热量穿过液态的外核，造成其内流体的翻腾。随着铁元素或铁-镍合金的结晶析出，熔融物中的杂质（大部分是熔解的硅酸盐类）也被分离出来。这部分物质的密度低于熔融态的外核，因而以连绵的颗粒雨的形式穿过外核而上升，那些颗粒或许也就如沙子般大小。杂质可能像上下颠倒的沉淀物一样集聚在地幔的基部，堆积在上下颠倒的山谷和洼地里。地震波显示，地幔基部有一层速度非常缓慢的地层，这种向上的沉淀物即可解释这种现象。这些沙样沉淀物可以圈闭熔融态的铁，就像海洋沉积物圈闭水一样。通过将铁保持在其内，这一地层所提供的物质可以在磁性上将地核内所产生的磁场与固态地幔所产生的磁场予以匹配。如果部分这种物质以超级地柱的形式上升，促成了地表的溢流玄武岩，就可以解释为什么这类岩石中含有高浓度的金和铂等贵金属了。

磁力发电机

从地表上看，地球的磁场似乎可以由地核的一个大型永久性磁棒产生。但事实并非如此。那一定是一台发电机，而磁场是由外核里循环的熔融态铁的电流产生的。法拉第曾证明，如果手头有电导体，那么电流、磁场和运动三者中的任意两个均可产生第三个。这是电动机和发电机的工作原理。但地球没有外部的电气连接。电流和磁场在某种程度上都是由地核内的对流产生和维持的，这就是所谓的自维持发电机。但它必须以某种形式启动。在地球拥有自身的启动装置之前，启动或许要依靠来自太阳的磁场。

地表的磁场相对简单，但产生地表磁场的地核内电流则必然要复杂得多。人们提出了很多模型，其中一些诸如旋转导电圆盘等构想只具有纯粹的理论意义。就我们看到的磁场而言，最有说服力的模型用到了一系列柱形胞腔，其中每一个都包含螺旋式的循环，这种循环是由热对流和地球自转所产生的科氏力共同作用所生成的。地球磁场最奇怪的特征之一，是它会在不规则的时间间隔内（通常是数十万年）反转极性，我们将会在下一章对此进行更详细的讨论。除此之外，有时会在长达5000万年的时期没有一次反转。单个火山晶体内所圈闭的磁场强度的证据表明，在磁极未反转的时期（即超静磁期），磁场可能比如今更强一些。磁场并非与地球的自转轴精确对准。当前，磁场与地球自转轴的倾斜角度大约是11°，但它不会永久停留在那个角度。1665年，磁场几乎指向正北，然后又偏移开去，1823年的倾角为向西24°。计算机模型无法准确解释这种情况，但提出了发电机本身无序变动的可能。在大部分时间，地球的磁场与地幔相配合，效力下降，但有时会产生很强的效力，以至磁场翻转。至于这种翻转能在一夜间完成，还是要持续数千年时间，其间磁场或者胡乱移动或者彻底消失，现在都还不清楚。如果是后一种情况，则不管是对罗盘导航还是对整个地球上的生命都是个坏消息，因为如此一来，我们都会暴露在来自太空的更危险的辐射和粒子之下。

还有人试图通过实验，为地核内发生的一切建模。这并非易事，因为需要大量的导电流体以足够的速度循环来激发磁场。德国和拉脱维亚的科学家们在里加实现了这一目标，他们使用装在同心圆筒中的2立方米熔融钠。通过将钠以15米每秒的速度推入中柱，他们最终创造了一个自激式磁场。

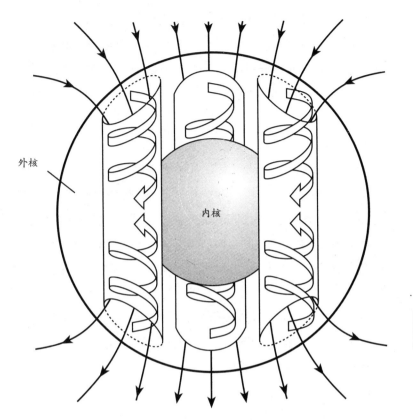

图7 地球磁场产生的一个可能的模型。外核中的对流因科氏力而呈螺旋状（带状箭头）。这与电流（未显示）一起产生了磁力线（黑色箭头）

给地球测温

地球越深处温度越高，但地球的中心究竟有多热？答案是，在地球的熔融态外核与固态内核的分界线上，温度必然恰好就是铁的熔点。但是，在那样的惊人压力下，铁的熔点会与其在地表上的数值大不相同。为了查明这个温度到底是多少，科学家们必须在实验室里重现那些条件,或是依据理论进行计算。他们尝试了两种不同的实用方法:一种是使用在金刚石砧间挤压的微小样本,另一种则是使用一台多级压缩气炮,瞬间压缩样本。因为很难达到如此巨大的压力——内核分界线处的330吉帕斯卡——并且很难校准压力以便知道是否到达目标,目前两种方法均未能直接测量出温度。它们能做的只是测量在压力略低的条件下铁的熔点,并以此为出发点向下推断。但还存在其他的难点,很大原因是由于地核并非纯铁质的,杂质会影响熔点。在纯铁状态下,理论计算得出的内核分界线温度是6500°C,而就地核可能掺杂的杂质多少而言,铁的熔点可能是5100—5500°C。这些数字都在通过金刚石砧和气炮这两种实验估算结果的范围之内。

对穿过内核的地震波进行的研究制造了另一个惊喜。这些地震波从北向南的行进速率似乎要比从东向西快3%—4%;内核表现出各向异性,一种在所有方向上各不相同的结构或晶粒。对此的可能解释是,内核是由很多对齐的铁晶体组成的——甚或是直径逾2000公里的一整块铁晶体!另一种可能是,内核内部的对流与地幔中的完全一样。或许有少量液体被晶粥所捕获。经过计算,与赤道对齐的3%—10%容量的液体平盘即可使内核呈现出科学家们观察到的各向异性现象。

旋转的地核

像整个地球一样,内核也在自转,但自转的方式与地球其他部分并不完全相同。它实际上转得比这个星球的其他部分稍微快一些,在过去30年里,走快了将近十分之一圈。在阿拉斯加探测到了来自南美洲最南端以外南桑威奇群岛的地震,对其地震波的仔细研究显示了上述结果。这是由上文刚刚讨论的内核的南北向各向异性所揭示的。因为内核超前于地球其他部分，该各向异性的结果发生了变化。1995年，堪堪掠过内核外界的地震波抵达阿拉斯加的速度与1967年的速度一样。但1995年地震波穿过内核的速度比1967年快了0.3秒，这表明内核的快速通道轴线每年向定线方向移动大约1.1°。理解了内核为何转得如此之快，我们便可以深入了解在那个强磁场环境里究竟发生了什么。可能的情况是外核中的电流对内核起到了一种磁性拖曳的作用，这与大气层中的射流相似。

到目前为止，整个地核中只有大约4%是冰冻的。但在30亿—40亿年后，整个地核都将凝固，届时，我们也许就会失去磁防护了。

第四章
海洋之下

隐秘的世界

在我们这个星球上,71%的表面积被水覆盖,其中只有1%是淡水,2%是冰,余下的97%都是海洋中的盐水。水体的平均深度是4000米,最深处可达11 000米。水面上露出的只是冰山一角。在深度超过50米,也就是所谓的透光带之下,阳光就很难穿透了。其下方是一个冰冷黑暗的世界,与我们的世界截然不同——至少在大约130年前的确如此。

1872年,英国皇家海军"挑战者"号启航,开始了海洋探索的首次科学远航。这艘船遍访了所有大洋,在四年中航行十万公里,但其航行的深度却只能通过在船侧放下一个砝码进行单点测量。因此,在第二次世界大战期间声呐和沉积物取芯等技术取得发展之前,海洋学进步的步伐还十分缓慢。冷战期间,西方诸国需要优质的海床地图,以便隐蔽自己的潜水艇,还需要先进的声呐和水下测声仪矩阵来侦测苏联的潜水艇。如今,船只安装和拖曳的声呐扫描仪已为大部分海床绘制了相当详细的地图。在很多地区,大洋钻探计划已经获得了其下岩石的样本,深水载人潜艇和潜水机器人也已经造访了很多有趣的处所,但亟待探索的空间仍然很大。

水从哪里来?

地球最初的大气层看来可能被初生太阳的太阳风刮走了大半。我们几乎可以肯定,促成地球最终形成的大爆炸以及产生了月球的大撞击此二者所产生的热量,必定熔融了地表的岩石,蒸发了大半原始水体。那么,如今我们看到的广阔大洋是从哪里来的?40亿年前最古老的岩石里埋藏着线索,它们在形成之时被液态水环绕着,水生细菌出现后不久也提供了有关的证据。印度在距今30亿年前的沉淀物中发现了最古老的雨滴化石印记。某些地表水可能是以火山气体的形式从地球内部逸出的,但大多数或许来自太空。时至今日,每年仍有大约三万吨水从遥远的外太空随着彗星粒子的细雨落向地球。在太阳系初期,水流量一定明显高得多,后来的很多撞击也可能是由整个彗星或其碎片所造成的,其组成被比拟为脏雪球,其中包含着大量的水冰。

咸味的海洋

如今,按重量计算,海水中大约有2.9%是溶解的盐类,其中大部分是食盐,即氯化钠;但也含有镁、钾、钙的硫酸盐、重碳酸盐和氯化物,此外还有些微量元素。海水中的盐度各不相同,取决于蒸发率及淡水的流入。因此,比方说,波罗的海的盐度较低,而被陆地包围的死海,其盐度大约是平均值(即每千克海水含35克固态盐)的6倍。但盐度中每种主要成分的相对比例在全世界范围内是一致的。

海洋并非一直这样咸。大部分盐类据信来自陆地上的岩石。一些盐类只是被雨水和河流溶解,而另一些则是由化学风化作用释放的,在风化过程中,溶解在雨水中的二氧化碳生成了弱性的碳酸。这种物质将岩石里的硅酸盐矿物缓慢地转变成黏土矿物。这些过程往往会保留钾而释放钠,这就是为什么氯化钠成为海盐中的最大组分。近数亿年来,海洋盐度大致稳定,风化作用与蒸发岩矿床和其他沉淀物的沉积作用二者所输入的盐分大致达到了平衡。

鲜活的海洋

海洋中还有很多微量化学物质,其中很多是对生命非常重要的养分,因而对于海洋生产力也至关重要。因此,它们往往在地表水体中便消耗了。将航行在太空中的彩色图像扫描器调试成对浮游植物的叶绿素等色素的特征波长敏感,即可绘制出海洋中季节性繁殖带的地图。最高的生产力往往发生在中高纬度地区的春季，在那里，温水与营养丰富的冷水相遇。1980年代，加州莫斯兰丁海洋实验室的约翰·马丁注意到，浮游生物的大量繁殖可能会在火山型洋岛周围造成下降流。他认为，铁或许是限制海洋生产力的一种养分，而火山岩会提供微量的溶解铁。从那以后，科学家们把小块的铁盐放置在南太平洋中，又观测了冰川期开始时海洋生产力最高地区的沉积岩芯，当时风吹来的尘埃为海洋提供了铁元素：这类实验和观测结果均证实了约翰·马丁的看法。但用铁来改善海洋生产力未必是应对日益严重的温室效应的良方，因为随着浮游生物的死亡或被食用，大多数被吸收的二氧化碳似乎又循环回到了溶液之中。

海洋的边缘

　　大陆的边缘往往会有一条窄窄的大陆架，深度只有200米。从地质角度来看，这事实上是大陆而非海洋的一部分，此外，在海平面低得多的时期，部分大陆架一定曾经是干燥的陆地。大陆架的生产力往往很高，渔业发达，至少在过度捕捞开始减少渔获之前如此。有机生产力与河流或风从附近大陆冲刷下来的大量粉砂、淤泥和沙子一起，积累了厚厚的沉淀物。在河流提供这些沉淀的地方，负载着沉淀物的浓稠水体有时会（几乎跟河流一样）继续流经峡谷和大陆架的边缘，有时则会继续流向大海达数千公里，直到最终散开，形成三角洲那样的地形——亚马孙河的情况就是如此。在一些地方，大陆架的边缘会有悬崖和峡谷等壮丽水下景观，虽然只有通过声呐才能看得到，但它们与陆地上的景观相比毫不逊色。

洋底

深邃广袤的洋底相对平整，平淡无奇，数英里范围内也不过偶然出现一些海参（它实际上是一种棘皮动物，是海星的亲戚），但那里也有山脉和峡谷。我们后面会提到洋中脊和海沟，但那里还有很多从洋底升起的孤立的海山脉，有时称其为平顶海山。这些平顶海山完全像是水下的山脉，往往是一些孤立的火山。它们是在过去由地幔柱生成的，不过并非位于构造板块的边缘。其中很多位于水下1000多米深处以下，但有证据表明它们曾经是升出海面的火山岛，被海浪侵蚀变平，又整个或部分地沉入深处。有时，下沉的速度慢得足以使珊瑚礁在岛周累积下来，在火山陆地消失之后，留下一个圈形的环礁。有时在洋底横越地幔柱时，会生成一串岛屿链。最著名的岛屿链组成了夏威夷群岛以及夏威夷西北部的天皇海山。

滑坡和海啸

大陆架和海山脉的陡峭边缘意味着那里的斜坡很容易变得不稳定。海床和大面积海底滑坡的周边海岸有证据表明，发生海底滑坡时，边坡坍塌使得数十立方公里的沉淀物像瀑布下落一般沉入深海平原。马德拉群岛和加那利群岛西面的大西洋、非洲西北部的外海，以及挪威北部的外海领域都存在着经过仔细研究的样本。有时，滑坡是由地震引起的，在其他情况下则只是沉积物堆积得太陡而导致斜面坍塌。无论是何种原因，水下的滑坡均可产生名为海啸的灾难性巨浪。有证据表明，过去3万年，挪威西北方向的挪威海曾经发生过3次特大的水下滑坡。其中的一次发生在大约7000年前，1700立方公里的碎石滑下大陆坡,冲向冰岛东面的深海平原。滑坡引发的海啸淹没了挪威部分陆地和苏格兰的部分海岸线,巨浪高达当时海平面以上10米。约10.5万年前,夏威夷的拉那伊岛南部曾发生过一次破坏性更大的滑坡。拉那伊岛经历了超过当时海平面360米的洪水,横越太平洋的海啸在澳大利亚东部堆积起的碎石高度达到海平面以上20米。

这些大滑坡,以及发生在大陆边缘的一些较为温和的小型滑坡所释放的沉淀物,被水体湍流抬升起来,可以散布到相当远的距离。滑坡产生了名为浊积岩的典型沉积物,其内的颗粒大小在不同湍流内逐级变化。初始的滑坡可能含有各种粒径的颗粒,但随着湍流成扇形展开,粗砂比细粉砂和淤泥流出得更快,因而各个流带会在其内对这些颗粒进行从粗到细的分级。如今,深水沉积岩层序中经常会发现这样的浊积岩。

海平面

我们的星球表面最明显的特质之一,就是陆地和海洋的分界线:海岸线。这是地球上变化最剧烈的环境之一,地貌呈现多样性,从高耸多岩的悬崖到低洼的沙丘和泥滩。另外不知为何,大量人群似乎特别喜欢在气候炎热的季节拥向此地。但海岸线并非一成不变。某些地段因为海水冲刷走数百万吨物质而遭到侵蚀。在其他一些地方,随着海水抬高沙洲或河流的泥泞三角洲扩大,陆地的面积不断增大。在地质时间尺度上,这些变化一度十分壮观。在某些事件中,所谓的"海侵"作用淹没了大块的大陆。而在其他时期海水撤退,这种现象被称为"海退"。海平面的这些明显变化可能是很多原因导致的。当前对于全球气候变暖的担忧之一,就是它可能会导致海平面上升。这可以简单地归因于海洋变暖导致水体稍微扩张，单单这一点，就可能在下个世纪将海平面抬升大约半米。如果南极冰盖发生明显的融化，海平面可能会升得更高。（北极冰和南极海冰的融化对海平面可能没有整体的影响，因为冰已经在漂浮，因而已经替代了其自身在水中的重量。）

但比起海平面在过去发生的变化，所有这些都不值一提。从上一次冰川期的高峰以来，海平面看来已经上升了160米之多。在过去300万年里，海平面在冰川期随着气候变化而剧烈变动。再回溯得更久远一些，在9500万到6700万年前之间的晚白垩世，海平面曾达到其最高位，那时的浅海覆盖了大陆的大片区域，产生了厚厚的白垩沉积，以及如今生产石油的许多沉积覆层。海平面如此异常升高，解释该现象的一个理论是，随着大西洋开始开放，洋底的大片区域也被地幔中升起的热物质抬升起来。海平面地质记录的特征就是在海洋的稳定上升期之后，海平面会出现明显的急剧下落。有时，海平面明显下落可以归因于大陆的构造隆起。在某些例子中，这种情况看来会发生在全球各地，且不一定发生在冰川期开始之时。有时，这或许是因为洋底突然大规模开裂，真正把洋底从海洋之下拉拽了出来。

海洋钻探

从1968年开始，美国领导的深海钻探计划使用一艘名为"格罗玛·挑战者"号的钻探船，以科学方法从洋底直接取样。该计划在1985年被国际性的"大洋钻探计划"所取代，后者使用的是改进的"联合果敢"号。项目进行了大约200个单独的航程或航段，每个历时两个月左右，在每一个区间分别钻探取得了岩芯样本。最深的钻孔超过两公里，总共采得数千公里长的岩芯样本。其中很多都包含不同深度的沉积物，最深可达火山玄武岩。它们都记录了自身的起源以及气候和海洋的变迁。沉积的速率非常缓慢，远远比不上侵蚀陆地与河流三角洲的速率。在高纬度地区，沉积物中还含有乘着冰山漂流的黏土和岩石碎片，冰山融化后，它们就被遗留在那些地方。在别处，乘风而来的沙漠尘埃和火山灰在深水沉积物中占据了更大的比例，有时还会伴随着微小陨石的尘埃、鲨鱼的牙齿，甚至还有鲸鱼的听小骨。

表层水体的海洋生产力很高，还经常会有各类浮游生物沉下的残余物。在相对较浅的水体，石灰质鞭毛虫和有孔虫类动

图8 海洋钻探船"联合果敢"号。塔架在水线上方60米处

物的石灰质骨骼随处可见，形成了石灰质软泥，固化后可形成白垩或石灰岩。但碳酸钙的溶解度随着深度和压力的增加而升高。在水中3.5—4.5公里深处，就到了所谓的碳补偿深度，在这一深度之下，微小的骨骼往往会溶解消失。在这里，它们会被硅质软泥取代，后者是由硅藻和放射虫的微小硅酸骨骼构成的。硅酸也会溶解，但在南大洋以及印度洋和太平洋的部分海域，未溶解的量也足以形成明显的覆层。在少数海域，通常是黑海等大洋环流受限之地，底层水没有氧气，黑色页岩沉积下来。这些黑色页岩有时富含未在厌氧条件下被氧化或消耗的有机物，这些物质可以慢慢变成石油。厌氧沉积物偶尔会散布得更加广泛，表现出所谓的缺氧事件，在那里，大洋环流的变化阻碍了富氧水体沉到洋底。

淤泥中的讯息

沉积岩芯承载着有关昔日气候的漫长而连续的记录。沉积物的类型可以揭示其周边陆地上曾经发生过什么——例如，有没有乘坐冰山漂流至此或是从沙漠乘风而来的物质。但钙质软泥中氧元素的稳定同位素的比率则保留了更加精确的记录。水分子中的氧以不同的稳定同位素的形式存在，主要是^{16}O和^{18}O。随着海水的蒸发，含有更轻的^{16}O的分子蒸发得更容易一些，使得海水富含^{18}O。除非有大量的水被锁在极地冰盖之中，否则富含^{18}O的海水很快会被降雨与河流再度稀释。因此，与间冰期的情况相比，此时被浮游生物吸收并堆积在沉积物中的碳酸盐会含有更多的^{18}O，沉积物中的氧同位素于是能够反映全球气候。通过将沉积物所记录的变化与2000多万年的时间相匹配，大洋钻探计划已经揭示出气候在这一时间尺度的波动，这似乎反映了米兰柯维奇循环，即地球轴线的游移不定，以及地球围绕太阳公转的偏心率。

1970年代，大洋钻探计划来到了地中海。那里的岩芯揭示的情况颇有轰动效应。有人向我展示了其中的一个，如今它保存在纽约哥伦比亚大学的拉蒙特-多尔蒂地质观测所。它由一层又一层的白色结晶物质组成，是一种盐类（氯化钠）和硬石膏（硫酸钙）的混合物。这些蒸发岩的岩层只可能是在地中海干涸的过程中形成的。甚至在如今，蒸发率仍然很高，以至于如果把直布罗陀海峡封堵起来，整个地中海海水会在大约1000年时间里蒸发殆尽。岩芯中数百米的蒸发岩意味着在500万—650万年前这段时间，这种情况一定发生过大约40次。钻探接近直布罗陀海峡时，科学家们遇到了一片卵石和碎片的杂乱混合物。这一定是大西洋突破直布罗陀海峡、重新灌满地中海时，世界上最大的瀑布形成的巨型瀑布潭。想象一下，当时海水轰鸣飞溅，该是怎样一番壮观景象。

在大洋钻探计划的近期航程中，天然气水合物覆层的钻探当属最有意思的项目之一。这是含有高浓缩甲烷冰的沉积物，是在深海洋底的低温高压条件下形成的固态形式。天然气水合物岩芯重回海洋表面则格外令人兴奋，因为它们很容易重新变回气体，有时还会引起爆炸。这种特性让研究变得有些困难，但据信它们的储量非常庞大，有可能在将来成为天然气来源，经济意义十分重大。有人认为，它们对过去突然发生的气候变化贡献不小。它们可能相当不稳定，一次地震便可将大量天然气水合物从洋底释放出来，升上水面，产生大量的气泡。海平面的突然下降也会让天然气水合物变得不稳定，导致强有力的温室气体甲烷的释放。5500万年前的那次全球气候突然变暖可能就是天然气水合物释放甲烷所导致的。有人甚至认为，近年来在子虚乌有的百慕大三角地区有船只失踪的报道即脱胎于大天然气泡打破水面的平静、弄翻船只或使船员窒息的描述。

大量有机物可以埋藏在海洋沉积物中，在合适的环境下，还能变成石油。这往往发生在正经历着地壳拉伸的浅海盆地。这一运动会把地壳拉薄，加深盆地，从而填补更多的沉积物。但与此同时，有机物被埋得更深，更接近地幔的内热，在这里被煮成了原油和天然气。随后，这些产物可以上升穿过渗透层，并聚集在不透水的黏土或盐层之下。岩盐特别容易移动，因为它的密度不太高，易于穿过大型穹地的地层。这些穹地常常会圈闭富油和天然气储备，墨西哥湾就是一例。

地下的生命

但是，海洋沉积物中的有机物并非都是无生命的。海底逾1000米以下的沉积物和上亿年历史的岩石中往往存活着大量有生命的细菌。它们似乎有可能在很久以前就被埋藏在海底的淤泥里，埋得越来越深，一直存活到现在。它们的生活算不上刺激，但也确实没死。据估计，它们可能每1000年才分裂一次，靠厌氧消化有机物而生存，并释放甲烷。某些细菌也能在或许高达100—150℃的高温下存活——这也是石油形成的温度范围，因而这些细菌可能在石油形成的过程中起到了重要的作用。所有的陆地细菌中可能有90%住在地下，它们共同组成了高达20%的地球总生物量。

地球上最长的山脉

如果从全世界的海洋中排干所有的水，让其下壮观的景色显露出来，你会看到那里最明显的特征并非比珠穆朗玛峰还高的巨大的洋岛山，也不是傲视美国大峡谷的宽广裂缝，而是一个长达7万公里的山脉：洋中脊系统。这些洋脊像网球上的接缝一样在地球上纵横。火山裂隙遍布整个山体。有时，这些裂缝在水下缓慢喷发，产生了枕状的黑色玄武岩熔岩凝块，像是挤出来的牙膏。这里是新生地带：随着海底的延展，新的洋壳在这里形成。

  北大西洋洋中脊是19世纪中期被一艘船发现的，当时该船正在铺设第一条横跨大西洋的电缆。洋脊宽广，宽度在1000公里到4000公里之间，缓慢升向中央的一列山峰，这些山峰的高度大都在洋底2500米以上，但距离海面仍有2500米。洋脊被为数众多的转换断层所断错，这些断层均垂直于其纵长，将脊顶移位达数十公里。脊顶常常由复线山峰组成，其间有一条中央裂谷。20世纪上半叶，阿瑟·霍姆斯等大陆漂移理论的支持者认为，洋脊或许标记了地幔中的对流将新地壳送到地表的位置，然而地磁测量最终证实了地质学最重要的发现之一：海底扩张。

磁化条带

  1950年代，美国海军需要洋底的详细地图来辅助潜水艇。于是，考察船开始往复航行，进行声呐测量。科学家们有机会进行其他实验，因而当时考察船拖着一台灵敏的磁力仪横越大洋，绘制了磁场的地图。该地图显示了一系列高低场强，像是分布在洋中脊两侧的平行条带。剑桥大学的弗雷德·瓦因和德拉蒙德·马修斯最终验证了实验结果。随着火山岩浆的喷发和冷却，它圈闭了与地球磁场对齐的磁性矿物颗粒。因此，航行经过全新世的海底玄武岩时，地球的磁场会增强些许。但是正如我们在上一章讨论的，地球的磁场有时会反转。在磁场反转期间喷发

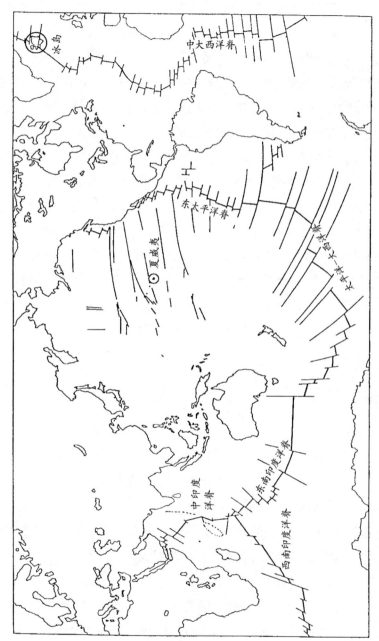

图9 洋中脊的全球体系以及将其切断的主要转换断裂带。图中用圆圈标记了夏威夷和冰岛等热点地区

出来的火山岩会圈闭与当时磁场相反的磁性,轻微降低磁力仪的读数。如此一来,洋中脊两侧的磁条带逐渐增加,从中脊向两侧移动得越远,其下的海底就越古老。海底的确在不断扩张。

新生的边界

总体而言,海底扩张的速度缓慢但从未停止,从太平洋的每年10厘米到大西洋的每年3—4厘米,大致相当于手指甲的生长速度。但岩浆喷发生成新地壳的速度并不稳定,这就是为什么部分洋脊在伸展的过程中会出现裂缝和凹陷,而其他洋脊却垒积成峰。在洋脊的中线之下,炙热的地幔物质以部分熔融的结晶

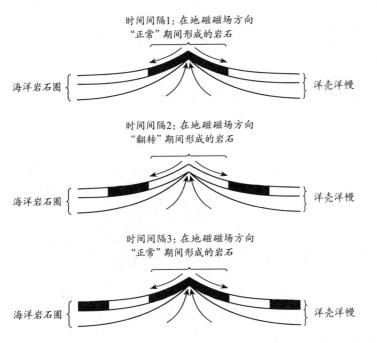

图10 随着新的洋壳从洋脊向两侧延展,洋底火山岩的磁化平行条带的形成过程

岩粥样物质的形式隆起。炙热柔软的软流层沿着这条线上升，遇到了一层薄薄的洋壳，其间并无任何坚硬的地幔岩石圈。因为这种地幔物质十分炙热，密度就比较低，因而使得洋脊上升。大约有4%的地幔岩熔融形成了玄武岩浆，向上渗透穿过气孔和裂隙，进入洋脊之下一公里左右的岩浆房。地震剖面图显示，太平洋部分洋脊之下的岩浆房有数公里宽，但大西洋洋脊之下的岩浆房却窄得难以觉察。岩浆房里的物质在缓慢冷却，所以有些物质结晶析出并积聚在岩浆房的底部，形成了一层质地粗糙的岩石，称作辉长岩。其余的熔融物定期从洋脊上的裂隙喷发出去。这些喷发物的流动性很强，且不含太多的气体或蒸汽，因而喷发的过程相当温和。但岩浆被海水迅速猝熄，往往会形成一系列枕状结构。

黑水喷口

即使没有活跃的火山喷发，靠近洋脊的岩石依然异常炙热。

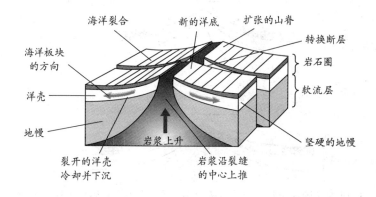

图11 洋中脊的主要成分

海水灌进干燥玄武岩的裂缝和气孔,在那里被加热并溶解硫化物等矿物。随后,热水从裂口处升出,硫化物沉淀形成了高耸中空的火山管。能够忍耐热水的细菌将可溶性硫酸盐分解为硫化物,也参与了这一过程。随着硫化物从这种不断冷却的水溶液中析出,它们就形成了一种黑色颗粒的云,因而这些出口通常被称为黑水喷口。水可以从中高速喷涌而出,温度超过350℃,因而在深潜潜水器中观察这场景既危险又令人着迷。矿物质火山管以每天数厘米的速度加长,直至崩塌成一堆碎片。如此一来,储量可观、潜在价值不菲的硫化物矿物便可堆积起来。在酸度更高但温度略低的水域,溶解的硫化锌更多,从而产生了白水喷口。这种喷口加长的速度更慢,通常温度也更低一些,因而对于聚集在这种热液喷溢口周围的某些神奇的生命形式而言,这里是更佳的栖息之所。这里的生命完全依赖化学能而非阳光。原始细菌在炎热且往往是酸性的环境中活跃生长。盲虾、盲蟹和巨蚌以它们为食,而体内含有共生细菌的巨型管虫从水中过滤养分。有人认为,地球上的生命最初便起始于这种地方,因此研究人员对它们很感兴趣。

来自海洋的财富

1870年代,"挑战者"号航行的惊人发现之一,就是它带回的那些奇怪的黑色结核块,那是来自深海洋底的挖掘样本。这些团块含有极其丰富的锰、铁氧化物和氢氧化物,以及具有潜在价值的铜、镍和钴等金属。这些团块被称为锰结核,如今人们知道它遍布深海洋底的大片地区。它们具体的成因尚不清楚,但那似乎是个漫长的化学过程,金属来自海水,还可能来自海底的沉积物。这些结核块往往围绕着一个很小的固体核心(也许是个玄武岩的碎片)、一团黏土,或者一颗鲨鱼牙齿,长成洋葱样的多层同心圆。对其地质年代的估计认为,它们生长得非常缓慢,大概100万年才会增长几个毫米。1970年代,人们提出了各种各样的方案,使用铲斗或抽吸的方式来开采这种矿物,但截至目前,由于技术、政治、生态和经济上的困难,采掘工作尚未开始。

推力、拉力,以及地柱

看起来海底扩张并不像是洋底从洋中脊系统推开来的结果。就大部分洋脊而言,其下并没有大量地幔柱热物质上升。看起来更像是洋脊被撕开,新的物质升起来填充缺口。洋脊下没有又厚又硬的岩石圈,只有几公里洋壳。地幔物质从洋脊下升起时,压力下降,某些矿物质的熔点也随之下降了。这导致有多达20%—25%的物质部分熔融,生成了玄武岩浆。岩浆形成的速度刚好能够生成厚度为7公里的相当均匀的洋壳。

冰岛是个值得注意的例外,那里的地幔柱和洋脊出现了重合。那里喷发的玄武岩远多于他处,地壳大约有25公里厚,这就是为什么冰岛高高耸立于大西洋之上。跟踪探测横穿格陵兰和苏格兰之间的北大西洋的加厚玄武岩洋壳,即可追溯那个地幔柱的历史。地震勘探显示,那里另有1000万立方公里的玄武岩,是阿尔卑斯山脉容积的数倍,足够以一公里厚度的地层覆盖整个美国。其中大部分洋壳并未喷发到地表,而是注入地壳之下,这一过程被称作"底侵"。格陵兰外海的哈顿滩正是玄武岩如此注入而导致的隆起。当前位于冰岛之下的地幔柱可能是导致北大西洋在大约5700万年前开始开放的原因。当时,火山活动似乎始于一系列火山爆发,其中某些火山迄今依然在苏格兰西北方向的内赫布里底群岛和法罗群岛保持活跃。

大洋死亡之所

　　洋壳一直不断地形成。其结果是，我们很难找到真正的远古洋底。最古老洋底的年代被确定在大约两亿年前的侏罗纪，地点位于西太平洋。人们最近在新西兰附近发现了一段洋底，大约有1.45亿年的历史。但这样古老的地质年龄很少见；大多数洋底的地质年龄不到一亿年。那么，远古海洋都去哪儿了？

　　答案就是一种被称作潜没作用的过程。随着大西洋拓宽，一侧的美洲与另一侧的非洲和欧洲缓慢分离。但地球总体上并没有变大，所以一定有什么东西在进行整合。这么做的似乎是太平洋。太平洋看上去被巨大的海沟所围绕，那些海沟最深可达11 000米。它们身后则是岛屿或大陆上的火山圈，也就是所谓的太平洋火山带。地震剖面测量显示了海洋板块——薄薄的洋壳及其下厚达100公里的地幔岩石圈——是如何重新陷入地球的。在其存在的一亿年期间，岩石圈的岩石不断冷却收缩，密度越来越大，以至于无法继续在软流层上漂浮。潜没作用的这一过程正是板块构造的驱动力之一：它是一种拉力而不是推力。

　　在潜没带下沉的冰冷致密的岩石已经到了海底，因而是湿的。气孔的空间里有水，矿物中也有化学结合水。随着板块下沉，压力和温度上升，水的存在为板块流动起到了润滑剂的作用，但也降低了某些组分的熔点，而这些组分通过周围的地壳上升，最终流入灼热的火山圈。正如我们在上一章看到的，岩石圈板块的其余部分继续流入地幔，至少流到了670公里深处，即上、下地幔的分界线上，但最终或许会下沉至地幔的基部。地震层析成像有助于跟踪其长达十亿年的行程。

在组成地球构造板块的大陆板块和海洋岩石圈之间,有几种不同类型的边界。在海洋中,有洋脊的建设性板块边界,还有潜没作用发生之处的破坏性板块边界。边界可能位于海洋岩石圈下潜到大陆以下之处,例如南美洲西岸形成了安第斯山脉的火山峰。海洋也可下潜到另一个海洋下面,就像西太平洋那些幽深的海沟,那里的火山圈组成了火山岛弧。在有些边界,一个板块会沿着另一个板块一路摩擦,比如加州沿岸。而在另一些板块边界,一块大陆撞击着另一块大陆,我们会在下一章讨论这种情况。

陆地上还剩下什么

海洋的消失并不会带走一切。在海洋岩石圈下潜到大陆下面,或是整个海洋在两大块陆地之间受到挤压的地方,很多沉积物被舀取上来,添加到大陆之上。这是为什么能在陆地上看到这么多海洋化石的原因之一。有时,整块洋壳也会被抬上陆地,这个过程叫作仰冲作用。因为来自碰撞地带,这类岩石往往是极度扭曲的,但把若干个这样的层序提供的证据拼凑起来,就能够一窥全豹。它们被称作"蛇绿岩层序"(ophiolite sequences),英文ophiolite源于希腊语,意为"蛇岩"。"蛇纹岩"这一名称形容的也是同样的事物,之所以有此叫法,是因为热水导致绿色矿物变形,看上去像在水中蠕动的线条。蛇绿岩层序的顶部是海洋沉积物的残余,其下是枕状熔岩和可能曾被注入地下的玄武岩层。接下来是辉长岩,一种缓慢冷却的结晶岩,其组成与玄武岩相同,在基部则是来自岩浆房底部的不同晶体覆层。在那之下可能有地幔岩的痕迹,玄武岩就是源自那里。

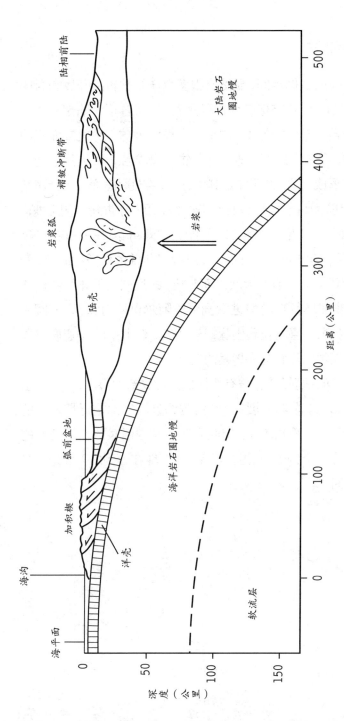

图12 海洋岩石圈如何潜没在大陆之下,边缘地区的沉积物不断累积,造成内陆的火山活动

失踪的海洋

在过去的数亿年间，显然有很多海洋都曾历经开放和封闭的过程。从12亿到7.5亿年前这段漫长时期，各个大陆聚集成一个巨型的超大陆，围绕着它的是一个横跨地球三分之二面积的广阔海洋。在前寒武纪后期，超大陆分裂成若干块大片陆地。新的海洋形成了。古大西洋是其中之一，存在的时间大约是6亿到4.2亿年前。其接合点，或称地缝合线，即海洋重新封闭之处，就位于如今的苏格兰西北部，短程驱车即可穿过——在5亿年前，那可要穿越5000公里的海面。在2亿年前的侏罗纪时期，西欧和东南亚之间曾有一大块洋面开放，即通向太平洋的特提斯洋。后来非洲环行移动撞进欧洲形成阿尔卑斯山脉、印度闯入中国西藏抬升了喜马拉雅山脉，这一海域又封闭了。地震研究找到了特提斯洋的洋底沉入地幔的残余物质。

在漫长的地质时期，曾有无数机会形成新的海洋，而实际上却都没有形成。东非大裂谷、红海与约旦河谷都是近期发生的明显例子。产生了北海石油储藏和巴伐利亚温泉的北海盆地的延展则是另一个例子。再过数亿年，我们现在看到的大洋地图又会完全过时了。

第五章
漂移的大陆

　　我小时候喜欢帮妈妈做橘子果酱。我承认,我现在还喜欢干这个,偶尔还会自己做。但如今每当我看着炖煮水果和糖的果酱锅,都禁不住觉得自己正在观望着我们所在的星球的演化,只是过程大大加速了，1秒钟大概相当于1000万甚至1亿年。果酱在小火上慢炖时会建立起对流圈，热气腾腾的橘子果酱团上升到表面，再四下散开。随之而来的是一些浮渣，那是些细小的糖沫，因为密度不够，无法再沉下去，只能聚集成片，漂在表面比较平静的地方。这些糖沫有点像地球上的大陆。它在整个过程的早期就开始形成，慢慢聚积变厚。其下的对流模式偶尔会发生变化，糖沫便分裂开来。有时候，浮沫会聚在一起，堆得更厚。当然，这样的类比应该适可而止。两者的时间尺度和化学作用都太不一样了；地质学家基本上不会在花岗岩中找到糖结晶，也不会在玄武岩中找到橘皮捕房岩。但在考察地球的糖沫——大陆时，我们不妨把这个形象记在脑子里。

地球的糖沫

陆壳与海洋底部的地壳大不相同。洋壳的主要组成是硅酸镁，而陆壳中含有更高比例的铝硅酸盐。相对于地幔或洋底所含的密度更大的物质，陆壳中所含的铁比较少。这就是为什么陆壳能够漂浮，尽管它是在半固态的地幔之上，而非在液体中漂浮。陆壳还可能很厚。洋壳是相当均匀的7公里厚，但陆壳的厚度可达30—60公里甚至更厚。此外，像海洋岩石圈一样，陆壳之下也有厚厚一层冰冷坚硬的地幔。大陆的根基到底有多深至今仍是学界争论的焦点，该争论很可能会以界定性的结论告终。但大陆又有点像冰山：我们看到的不过是浮出水面的一角。大陆表面的山脉上升得越高，基本上它的根基也就下扎得越深。

漂移的大陆

得益于后见之明、有关地幔对流的知识和海底扩张的证据，我们很容易看到,在漫长的地质时期,大陆相对于彼此曾有过位移。但并非所有的证据都令人信服。虽然詹姆斯·赫顿提出了关于造山运动和岩石循环的理论,但任何原理的提出都需要很长时间。1910—1915年间,美国冰川学家弗兰克·泰勒[①]和德国气象学家阿尔弗雷德·魏格纳提出了大陆漂移的假说。但在当时,没有人能够想象大陆如何像船一样在看似固态的石质地幔上漂流。其后近半个世纪,大陆漂移的支持者一直是少数。然而该学说的少数支持者非常勤奋。南非的亚历克斯·杜托伊特积累了南非和南美洲之间类似的岩石结构的证据,英国地球物理学家阿瑟·霍姆斯则提出了地幔对流作为漂移的原理。直到1960年代,海洋学家们着手这项工作时,辩论才尘埃落定。哈里·赫斯指出,洋壳下的对流可能导致了海底从洋中脊向外扩张,弗雷德·瓦因和德拉姆·马修斯也提出了海底扩张的地磁证据。多亏有加拿大的图佐·威尔逊、普林斯顿大学的杰森·摩根和剑桥大学的丹·麦肯齐等人发表的论文,才将各方面的证据拼接起来,形成了板块构造学说。

　　板块构造用少量坚硬板块之间的相对位移、相互作用及其边缘的变形,对地球表面做出了解释。这并不是说大陆在自由漂移,而是它们被架在板块上,这些板块延展得很深,包括地幔岩石圈在内,通常厚达100公里。板块并不限于大陆,还包括洋底的板块。地球上一共有七个主要的板块:非洲、欧亚、北美、南美、太平洋、印澳,以及南极洲板块。还有一些小板块,包括环

---

① 弗兰克·泰勒(1860—1938),美国地质学家。从哈佛大学辍学后,在父亲的资助下成为北美五大湖地区的冰川学专家。1908年底向美国地质学会投稿,提出了大陆漂移的假说,但遭到其他科学家的忽视或反对。

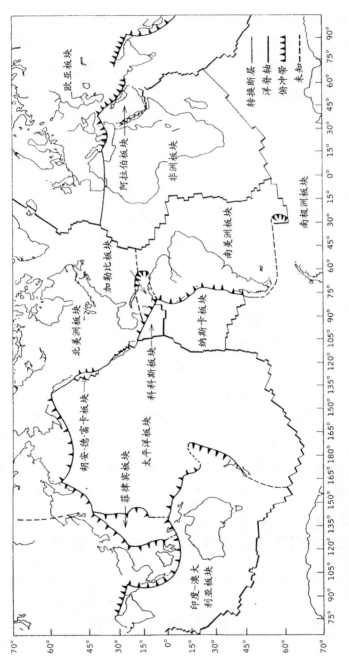

图 13 世界的主要构造板块及其边界

绕太平洋的三个相当坚固的小板块，以及板块接合之处的一些较为复杂的碎片。

我的另一个童年记忆是在世界地图上寻找大陆，把它们剪下来，并试着把它们拼成一块大片的陆地。那一定发生在1965年图佐·威尔逊在《自然》期刊上发表论文那段时间前后。我还记得自己当时激动地发现这些大陆彼此相当吻合，并找到一些原因说明它们为何无法完美吻合。倒不是因为我剪得不够精细。每一个小书呆子都知道，应该沿着大陆架的边缘而不是沿着海岸线剪下大陆。可以切掉亚马孙河三角洲，否则它会与非洲重合，因为自从大陆分离之后那里又有了新的进展。更让人兴奋的发现是，北美洲和南美洲需要分开才能完美接合，西班牙则必须和法国分手。如果把西班牙转回去，就会在如今的比利牛斯山脉那里撞上法国。那么，这种大陆碰撞是否就是造成山脉的原因呢？

大概是在我青春期的时候，全家假期旅行时带我去了比利牛斯山和阿尔卑斯山。在一些地方，我看见那里的沉积岩层并不像其他受干扰较少的地区那样平整，而是皱巴巴的，像是叠起来的起伏不平的地毯。这把我的思绪又带回到橘子果酱那儿。炖果酱时要把一只瓷盘子放在冰箱里。每隔几分钟把它拿出来，在上面滴上几滴滚热的果酱。果酱冷却后，把手指按上去。如果果酱还是液体，那就舔舔手指头在一边等着，让它接着炖。但是过一会儿，果酱接近其凝固点时，放在盘子上的样品用手指一按就会起皱，就像大陆碰撞的缩微模型。对于巨大尺度上的大陆行为而言，这可是个不赖的模型。岩石重叠受到了难以置信的压力，还可能从下方被加热，在大陆碰撞的侧向力的影响下，会倾向于折叠而非破碎。受影响的巨石会受到重力的强大作用，因此，在自身重量的作用下，最险峻的褶皱会下垂成为过度褶皱，看上去就越发像软质奶油冰激凌或橘子果酱的外皮了。

地球不是平的

从地图上剪下来的平面大陆不会彼此完美吻合的另一个原因是，它们所代表的应当是球体表面上的板块。在投影地图上，它们被扭曲了。但在球体表面滑动坚硬的板块也并非易事。简单的直线移动显然不行，因为球体上没有直线。每一点移动实际上都是沿着穿过球体的轴进行的转动。但还有其他难题。其一是在所有相互碰撞板块中找到一个参照物，其二是要将海底扩张的不同速率考虑在内。要研究大西洋的开放以及美洲从非洲和欧洲分离的相对位移，简单模型可能会调用一个类似地球自转轴的轴线。但那就要造出橘皮一样的大西洋洋壳，赤道位置较宽，朝着两极的方向均匀变窄。海底扩张的速率的确各不相同，但这一点无法直观展现。扩张的结果就是转换断层；这些数千公里长的地壳断裂有效调整了洋中脊片段的偏差。

参照系

有了海底扩张的证据和地幔对流的原理，板块构造学说迅速建立起来，成为现代地球科学的核心理论。但直至现在，仍有地质学家反对使用"大陆漂移"这一术语，因为它会让人联想到这一原理尚未得到正确解释且几乎无人信服的那段时间。然而，一旦人们准备好接受这一理论，过去的板块运动的证据便明确显现了。地质学证据可以证明同种类型的岩石分开后，如今位于一个大洋的相反两侧。现存生物和化石遗迹的证据亦可证明在过去的不同时期，不同的生物种群彼此隔离，有时还能够在大陆间穿越。例如，澳大利亚与亚洲的某些部分，如马来西亚和印度支那，分开至今也不过两亿年。从那时开始，这两片大块陆地上的哺乳动物各自独立演化，结果是有袋类动物在澳大利亚占据了主要地位，而有胎盘的哺乳动物在亚洲得以发展。

  我们在上一章讨论过洋底玄武岩中有地磁反转的证据，同样，地磁证据也提供了昔日大陆运动最全面的画面。在火山岩凝固时像微小的罗盘指针一样被圈闭其中的磁性矿物颗粒，记录着当时北极的方向。它们不仅显示了磁场本身的小摆动和大逆转，还描绘出千万年甚至数亿年间更广范围内的大型曲线系列，即所谓的磁极迁移曲线。这实际上正是大陆本身相对于磁极如何位移的图示。在比较不同大陆的曲线时，有时会看到它们一起移动，有时又分道扬镳，看到各个大陆本身分离、漂流，又重聚，翩然跳出一支大陆圆舞曲。实际上，这更像是一支笨拙的谷仓舞，因为大陆步履凌乱，有时还会撞在一起。

  灵敏的仪器甚至可以跟踪当今大陆的相对位移。在短距离内，比如局部跨越板块边界的地方，测量技术可以做到非常精确，尤其是激光测距。但如今人们也可以通过太空完成整个大陆级别的测量。某些有史以来发射的最奇特的空间卫星就是用作激光测距的。这种卫星有一个由钛等致密金属组成的球体，上面镶嵌了很多猫眼一样的玻璃反射器。这些反射器能够原路反射向其射来的光线，因此，如果从地面射出一束精密的强力激光，并原地记录反射脉冲的时间，就可以算出精确到厘米的距离了。比较来自不同大陆的数值，就能看到年复一年，它们发生了怎样的位移。利用来自遥远宇宙的无线电波作为参照系，天文学家可以用射电望远镜进行同样的测量。现在，美国军方GPS（全球定位系统）卫星已经取消扰频设置，地质学家在野外使用小型手持GPS接收器即可达到相似的精度。细致使用多种读数的话，精度甚至可以提高到毫米级。答数证实了海底扩张速度的证据：板块相对位移的速度大致相当于手指甲的生长速度，即每年大约3—10厘米。

当然，所有这些板块运动的测量都是相对的，人们很难为所有的测量建立一个整体的参照系。夏威夷出现了一个线索。夏威夷的大岛[①]是一系列火山洋岛中最后出现的，这些火山洋岛向西北方向延展，继而潜入水下，形成天皇海山链。玄武岩的地质日期显示，越往西北方向去，当地的玄武岩就越古老。这个岛链似乎标记着一个通道，太平洋板块经这里横越其下一个充满地幔热物质的地柱。将此地幔柱的历史位置与其他地幔柱加以比较，就会看到它们之间几乎没有什么相对位移；如此一来，这类地幔柱或许可以作为基准点，因为其下的地幔几乎没有过什么变化。至于以此为参照系的绝对板块运动，对它们的估计显示，西太平洋移动得最多。与之相反，欧亚板块几乎没有移动，因此，历史上选择格林尼治作为经度的基准点，或许有其合理的地质学依据！

大陆圆舞曲

将地质学和古地磁学的证据结合起来，就能通过地质时间来回溯构造板块的运动。我们如今看到的大陆是由一块超大陆分裂而来的，这块超大陆被命名为联合古陆。在大约两亿年前的二叠纪时期，超大陆分崩离析，起初形成了两个大陆：北面的劳亚古陆和南面的冈瓦纳古陆。那些大陆的分裂至今仍在继

---

① 即夏威夷岛，夏威夷群岛中最大的岛屿，位于群岛最南端，面积10 414平方公里。

续。但回溯到更加久远的过去，联合古陆本身似乎是由更早时期的多个大陆集聚形成的，再向前回溯，曾经存在过一个更为古老的超大陆，名叫潘诺西亚大陆，而在它之前的超大陆叫罗迪尼亚大陆。这些超大陆分裂、漂移再重新集聚的周期被称为威尔逊周期，得名于图佐·威尔逊。

再回溯到前寒武纪，时间越久远，画面就变得越模糊，也就越难辨认出我们如今所知的大块陆地。例如，在大约4.5亿年前的奥陶纪时期，西伯利亚靠近赤道，大块陆地大多聚集在南半球，现在的撒哈拉沙漠当时则靠近南极。在前寒武纪后期，格陵兰和西伯利亚在距离赤道很远的南方，亚马孙古陆几乎就在南极，而澳大利亚却完全位于北半球。

就如此长距离的漂移而言，当前的纪录保持者之一显然是亚历山大岩层这块陆地。它现在形成了阿拉斯加狭地的大部分。大约5亿年前，它曾经是东澳大利亚的一部分。岩石中的古地磁学证据包括水平面沉向地球的倾斜度，这显示了岩石形成之时的纬度。倾角越小，纬度越高。其他线索来自锆石矿物的微小颗粒。它们携带着放射性衰变的产物，后者记录了它们形成之时的构造活动时期。就亚历山大岩层来说，它们显示了两个主要的造山期，分别是5.2亿年前和4.3亿年前。在这两个时期，东澳大利亚都是造山的发生地，而北美洲却一派平静。相反，北美洲西部在3.5亿年前十分活跃，那时的亚历山大岩层却似乎处在休眠状态。3.75亿年前，亚历山大岩层开始从澳大利亚分离出来，形成了一个水下的海洋高原，彼时有各种海洋动物在那里形成化石。

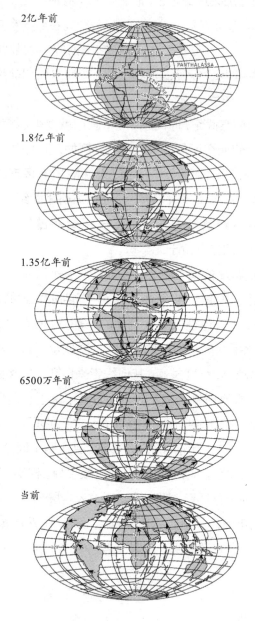

图14 2亿年来的地球大陆变迁图

大约2.25亿年前,亚历山大岩层开始以每年10厘米的速度向北方移动。这一过程持续了1.35亿年,在那段时间,随着亚历山大岩层抵达当前的纬度并与阿拉斯加发生碰撞,北美洲的化石也开始出现了。甚至还有可能该岩层在行程中与加州海岸擦肩而过,从加州马瑟洛德淘金带刮掉了一些物质。如果情况果真如此,那么阿拉斯加淘金潮跟加州淘金潮赖以采掘的或许都是同样的岩石,只不过位置北移了2400公里而已。

大陆的堆叠

我们听说过几种不同类型的板块边界。有洋中脊的扩张中心、与洋脊垂直的转换断层,以及海洋岩石圈插入大陆下方的潜没带等等。所有这些都是相对狭窄、界定清晰的地带,用简单图表便可相对容易地理解和解释。但还有一种更加复杂的板块边界,坚硬板块的构造学说无法解释这一现象,那就是陆内碰撞。涉及洋壳的则相对简单。只要洋壳够冷,其密度就足够致密,保证其以相对陡峭的约45°角沉入地幔。陆壳不会下沉,它们像在海中漂浮的软木塞一样始终保持漂浮状态,无惧拍打过来的惊涛骇浪。陆壳也比海洋岩石圈更容易变形。因此,当大陆碰撞时,那情形更像是一场严重的交通事故。

印度接合亚洲的过程就是一个很好的例子。在那以前的数亿年,印度一直在南半球参演一场复杂的土风舞,在场的其他舞者包括非洲、澳大利亚和南极洲。其后,大约1.8亿年前,印度分离出来,开始向北方漂移。印度西侧有一些非常壮观的山脉,即西高止山脉和德干地盾。这些山脉共有一个奇怪的特征:尽管海洋就在西面的不远处,但主要的大河仍然把这些山脉中的水排向东方。巴黎大学的文森特·库尔蒂耶教授开始研究组成这些山峦的玄武岩中的古地磁时，还遭遇了另一个谜题。他曾在喜马拉雅山脉研究经年，希望能与更南方的样本进行比较。他本以为会在厚厚的玄武岩地层中找到历时数百万年、跨越多个古地磁反转的古地磁数据。但事与愿违，他发现那些岩石的磁极都是同向的，这表明它们一定是在最多100万年这样一个短暂时期内喷发出来的。牛津大学的基思·考克斯和剑桥大学的丹·麦肯齐对其间的过程得出了较为肯定的结论。印度曾经是个体量更大的大陆，在向北漂移的路上，它穿越了一个当时正值岩浆喷发期的地幔柱。这导致印度大陆的隆起。但德干地盾刚好在这块穹地的东侧。西侧是上坡，河流无法西去。难以想象的火山喷发在数千年时间里生成了数百万立方公里的玄武岩。最终，火山活动把大陆一分为二。我们如今知道的印度次大陆只是原大陆的东北部分。海下的其余部分则躺在塞舌尔群岛与科摩罗群岛之间巨大的玄武岩积层上。事实上，这一火山流泻发生在大约6500万年前的白垩纪／三叠纪分界线前后，包括恐龙在内的很多动物种群也是在此期间灭绝的。也许杀死它们的根本不是某颗小行星，而是这些惊人的火山喷发所导致的污染和气候变化。

在此期间，次大陆的其余部分继续北移，封闭了特提斯洋的巨大洋湾，最终撞进亚洲。虽说特提斯洋的海洋岩石圈质地致密，可以潜没进亚洲下面的地幔，但陆壳却不能如此。这两个大陆的首次接触是在大约5500万年前，但其中一个大陆以巨大的动能在其轨道上不顾一切地前行，简直没有什么可以阻止。封闭的速度大约为每年10厘米，后来逐渐降到了每年5厘米左右，但从那时至今仍在继续碰撞，就像电视上慢速回放的车辆撞毁试验。在此期间，印度次大陆又继续向北行进了2000公里。首先发生的是沉积物的堆叠，以及随着印度大陆板块楔入亚洲之下，一系列潜没断层的地壳随之变厚，像是在推土机前垒起一堆碎石。这层厚厚的大陆物质使得喜马拉雅山脉至今仍在上升。

向北穿越平坦的恒河平原，我们能看到第一个巨型逆断层形成了当地的一个明显特征。山峦上的沉积物随处可见，这些沉积物曾经淤积在河床上，但现在被抬升了数十米，不过它们只有

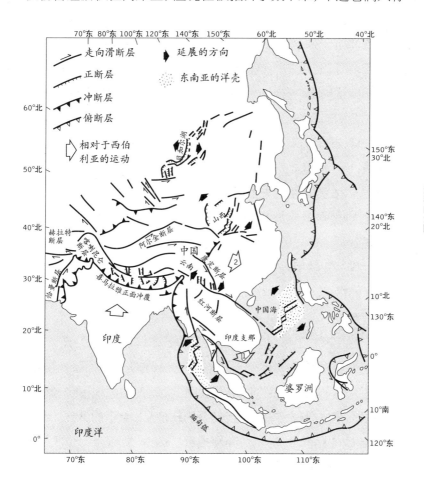

图15 东南亚构造图，显示了印度次大陆碰撞以及中国和印度支那被挤向一侧的运动所导致的主要断裂

几千年的地质年龄，表明这样大幅度的抬升是在数次地震中突然发生的。这些山峦是喜马拉雅山脉的山麓小丘，喜马拉雅山脉在一系列东西走向的山脉中崛起。每一条山峰线都大致对应着另一个大陆岩石的巨型楔入。如今暴露在喜马拉雅山脉表面的岩石大都是从大陆深处抬升上来的远古花岗岩和变质岩。从特提斯洋的洋底提升上来的褶皱沉积物则长眠在西藏高原边缘山脉的北方。在它们之后是一系列湖泊，它们对应着两块大陆的初始接合处，也就是所谓的地缝合线。

印度这样一个古老冰冷的大陆非常结实，而且相当坚硬。它所碰撞的亚洲部分则相对年轻和柔软。就像炙热的地幔可以经历固体流一样，地壳岩石也可如此。我们在地表看到岩石时，认为它既硬又脆，但石英等地壳矿物在仅仅数百摄氏度条件下就可以像太妃糖一样流动，地幔更深处的橄榄石也是一样。如果将印度看作相对坚固的大陆，与具有很多液体特性的另一方相撞，就能得到一些印度碰撞亚洲的最佳模型了。要说这另一方像液体，不如说它像半凝固的颜料——越用力挤压，它就越容易变形。这类模型可以解释中亚山脉的格局，却无法解释西藏高原的情形。

中国西藏的崛起

密度相对较低的地壳岩石的堆叠，无法只靠向下的逆冲推覆作用来调节，因而整个区域开始向上浮动。西藏地下致密的岩石圈根部与之分离，并沉回到其下的软流层。余下的日益加厚的大陆岩石向上浮动，将西藏高原抬升了8公里之高。与此同时，亚洲的一部分试图滑行让路，印度支那滑到了东面。这一侧向动作把大陆朝着更远的北方延展，造就了很多地貌特征，其中就包括西藏地区充满湖水的裂口和俄罗斯境内贝加尔湖的深纵裂口。随着地下冰冷致密的岩石圈部分移开,炙热的软流层距离西藏地壳就更近了,足以导致局部的熔融,也解释了西藏部分地区近来发现的火山岩。另有地震学证据表明,西藏高原西南部地下约20公里处有一个部分熔融的庞大花岗岩池。这也有助于解释亚洲如何吸收了来自印度的冲击,以及尽管西藏高原被高耸的山脉围绕,为何它本身还能保持相对平缓。总之,西藏高原看来不大可能会比当前平均5000米海拔更高了。再有任何抬升都会被物质向两侧流动的动作所抵消。多山地区的平均海拔也多半不会超过5000米。在这里,高度被侵蚀所制约。尽管喜马拉雅山脉已经有大量物质被侵蚀,巴基斯坦北部的南伽帕尔巴特峰等地区如今仍以每年若干毫米的速度上升,致使山坡变得崚嶒崎岖,并容易发生滑坡。

季风

喜马拉雅山脉的高度几乎相当于客机正常的飞行高度,这些山脉对大气循环构成了明显的障碍。因此,北面的中亚地区冬季寒冷,且全年大部分时间气候干燥。夏季,西藏高原升起的暖空气阻止了来自西南方的潮湿空气,因而云层堆积,并在印度季风的暴雨中释放其水分。季风在阿拉伯海兴风作浪,把养分带到表层水,从而导致浮游生物一年一度的繁殖期,继而在水下的沉积物中留下了痕迹。沉积物岩芯显示,这一循环始于大约800万年前,或许正对应着西藏高原大抬升的末期和季风气候形态的源起。在中国,乘风而来的尘埃显示,喜马拉雅山脉北部地区也是在这一时期才开始变得干燥。非洲西海岸之外的沉积物也有变化,随风而来的尘埃在地层中有所增加;与其相对应的似乎是非洲开始变得干燥,以及随着潮湿的云层被拉向印度,撒哈拉沙漠开始形成。有一种理论认为,在喜马拉雅山脉被侵蚀的过程中必定发生了大量的化学风化作用，消耗了大气层中的大量二氧化碳，这可能为过去2500万年各次冰川期的出场做好了铺垫。这么说来，我们知道是非洲的气候变化造就了物种演化的压力，导致现代人类在那里得到发展，或许那次气候变化的原因之一就是西藏和喜马拉雅山脉的崛起。

瑞士蛋卷

从喜马拉雅山脉的大陆堆叠再向西，特提斯洋收窄成一个水湾，但在意大利和非洲板块与欧洲碰撞的例子中，碰撞的结果差别不大，只不过后者是略小一些的版本而已。阿尔卑斯山脉是学界研究最多、理解最深的山脉之一。在阿尔卑斯山脉北面有一个沉积盆地，其中慢慢地填满了名为磨砾层的沉积物。山脉南面的意大利境内有一个波河平原，相当于印度的恒河平原。波河平原与山脉之间是一系列的沉积楔，这种沉积物叫作复理石，是由特提斯洋舀取而来的。随后我们来到了瑞士境内高耸的阿尔卑斯山，它由大陆的结晶质基部以及下方部分熔融的花岗岩侵入体共同组成。越过高山，会看到一系列强烈褶皱的岩石，这些岩石被舀取出来，形成巨大的倒转褶曲，名为推覆体，它们折向北方，因为自身的重量而下垂，像搅打过的奶油被舀起来那样。这些推覆褶皱往往延展得很远，以至于地质年龄更加古老的岩石会堆叠在年轻岩石的上方，形成令人非常困惑的序列。与喜马拉雅山脉一样，那里也有一系列逆断层，在很多地方使得大陆地壳的厚度加倍。

克拉通

没有哪个大陆是孑然孤立的岛屿。各个大陆既可分离，也可连接合并。这一结论已屡经验证，阿尔卑斯山脉和喜马拉雅山脉等现代山脉只是最近发生的几个实例而已。其他山脉因为过于古老，经年累月，几乎已被夷为平地。苏格兰西北的加里东山脉和北美洲的阿巴拉契亚山脉也可作为例证，只不过时间要追溯到大约4.2亿年前，大西洋的某个先驱封闭之时。现代大陆都是处处显现出这种特征的百衲被。然而随着大陆变得越古老越厚重，它就会越坚硬，也越持久。受构造运动影响最小、最稳定的大陆核心被称作克拉通，它们组成了当今南北美洲、澳大利亚、俄罗斯、斯堪的纳维亚和非洲的核心。久而久之，它们也常常会经历缓慢的下沉。澳大利亚的艾尔湖和北美洲的五大湖就位于这样的盆地。与之相反，南部非洲的克拉通却已被其下地幔柱的浮岩抬升了。

大陆的剖面

地震层析成像揭示了整个地球的结构，应用同样的原理也可以巨细靡遗地研究大陆的内部深处。为得到所需的高分辨率，这一技术并不依赖地球另一侧发生的随机自然地震（需要间隔很远的测震仪才能探测到它们），而是自行产生人工地震波，利用附近间隔紧密的探测仪矩阵来拾取反射波。地震层析成像造价昂贵，起初被石油勘探公司所垄断，这些公司小心翼翼地守护着探测结果。但如今很多国家项目都在共享这类数据。其中最先进的当属北美洲，美国的大陆深度反射剖面协会和加拿大的岩石圈探测计划均已建立起详细的剖面图系列。为了产生震波，他们使用一个小型的专用卡车车队，利用水锤泵带动重型金属板来震动地面。绵延数英里的传感器网络监测这些深度震动，并记录来自地下许多地层的反射。计算机分析可以显示出每一个不连续面或密度突变。这些剖面图远比石油勘探者最感兴趣的沉积盆地还要深得多。它们显示了在很久以前合并的各个大陆之间的远古缝合线，还显示出沉降到加拿大苏必利尔湖地区下方地幔中去的一个地层的反射，它很可能是迄今发现的最古老的潜没带，其洋底属于一片早已消失的海洋，地质年龄大约为27亿年。这些剖面图显示出从地幔上升却无法穿透厚重大陆的玄武岩岩浆如何在大陆之下铺垫玄武岩石板，即所谓的岩墙；还显示了当大陆岩石被埋得足够深时，它们是如何开始熔融，从而上升穿过大陆，重新结晶成为花岗岩的。

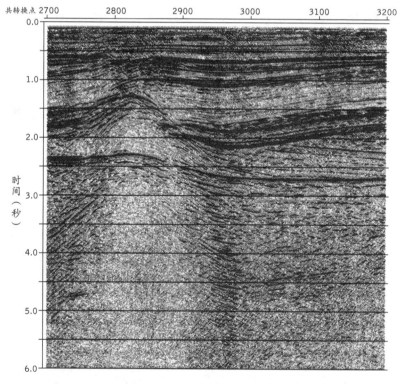

图16 地球地壳内的各地层和一个穹地结构的地震反射剖面图示例

花岗岩的上升

随着大陆岩石的堆叠，大陆的基底也被越埋越深。大陆下沉时被加热，基底的岩石开始熔融。这些岩石中很多都是数十亿年前沉积在海底的远古沉积物。它们含有与岩石化学结合的水。水有助于岩石的熔融，并有润滑作用，使它们容易上升到表面。与火山岩不同，它们过厚过黏，无法从火山中喷发出去。相反，熔融岩石的巨大气泡或达数万米之巨，向上推进更高层的大陆地层，速度也许相当快。它们炙烤着周围的岩石，缓慢冷却后形成由石英、长石和云母组成的粗糙结晶岩：花岗岩。最终，周围的岩石逐渐磨损，露出壮观的花岗岩穹隆，达特穆尔高地①即是一例。

对一个由硅酸盐岩石和大量的水组成的、构造运动活跃的星球而言，花岗岩的形成或许是不可避免的。但绝不会出现一个没有大陆而只有环球海洋的"水世界"。只要有水，它就会设法参与岩石的化学过程，在它们熔融之时提供润滑，以便它们能够以大块花岗岩的形态上升，形成大陆的顶峰高出海面。如果没有水，那就是金星上的情形了：没有板块的大地构造。如果没有熔融岩浆的内火，就是火星上的情形：古老冰冷的地表，即使有生命，也深藏在地下。在地球上，海洋和大陆始终处于动态的相互作用中，有时这种相互作用是你死我活的。

地球中的宝藏

地质勘探最初的动机之一便是寻找矿藏。若干地质过程可以形成或浓缩出珍贵而稀有的物质。地球的热量可以温和加热

---

① 位于英格兰西南部德文郡的一个高原。

沉积盆地的生物遗迹，从而生成煤炭、石油和天然气。我们已经看到了深海热液喷溢口周围何以富集贵金属硫化物，也看到了锰结核在深海洋底形成的过程。矿物以多种方式在大陆岩石中富集。在熔融的岩石中可以形成结晶，其中密度最大的会沉到熔体腔的底部。在聚集矿物质的同时，上升并穿过其他岩石的熔融岩石团块会驱使超热的水和蒸汽先行上升。在压力下，这种水汽混合物可以溶解多种矿物，特别是那些富含金属的矿物，并推动它们穿过裂缝，那些裂缝曾是它们作为矿脉的寄身之所。其他矿物会在水分蒸发或岩石中的其他组分受到侵蚀时在岩石表面浓缩。只要我们拥有使之再生的技术，地球中的宝藏就唾手可得了。

寻找失落的大陆

如果说在地球历史的大部分时期，大陆浮沫一直都聚集在地球表面，那它是从何时开始的？第一块大陆在哪里？这还很难说。最古老的大陆岩石经过改造、折叠、断裂、掩埋、部分熔融、再度折叠和断裂，被新生侵入岩渗透。此过程是如此漫长复杂，以至于很难清楚地解释，简直有点像从垃圾场的压实废料中辨认某一辆车的残骸。但寻找地球上最古老岩石的探索恐怕已经接近尾声。第一批竞争者中有些来自南非的巴伯顿绿岩带。这些岩石的地质年龄超过35亿年，但它们是枕状熔岩和洋岛的残留物，而不是大陆岩石。如今，人们在澳大利亚西部的皮尔巴拉地区找到了相似的岩石，格陵兰西南部也有距今37.5亿年的岩石，但这些也是海洋火山岩。第一块大陆的最佳候选者长眠在加拿大北部腹地。在耶洛奈夫镇[①]以北大约250公里无人居住的不毛之地，靠近阿卡斯塔河的位置，孤零零地立着一个工棚，里面装满了地质锤和野营装备。门上方有一个粗糙的标牌："阿卡斯塔市政厅，建于40亿年前。"附近某些岩石的地质年龄已逾40亿年。

那些岩石之所以泄露了自身年龄的秘密，要多亏在其晶格中圈闭了铀原子的锆石矿物颗粒，铀原子经过衰变，最终变成了铅。这些颗粒难免会因为再熔化、后期生长，以及宇宙射线损伤而受到影响，但澳大利亚研发的一种名为"高灵敏度高分辨率离子探针"的仪器，可以使用氧离子窄射束来轰击锆石的微小部分，以便对颗粒的不同区域单独进行分析。某些颗粒的中心部分显示其地质年龄高达40.55亿年，使得它们荣膺地球上最古老岩石的称号，并以自身的存在证明了地球形成不到5亿年即出现了大陆这一结论。

一粒沙中见永恒

但仍有诱人的证据表明，还有些历史更为久远的老古董。在澳大利亚西部，珀斯市以北大约800公里的杰克山区有砾岩岩石，这是大约30亿年前的圆形颗粒和小鹅卵石固结在岩石中的混合物。在岩石的颗粒中间有锆石，这一定是更早期的岩石受到侵蚀而析出的。其中一颗测出的地质年龄是44亿年，而对结晶中氧原子的分析表明，当时的地表一定冷得足以让液态水凝结。这一研究表明，有些大陆出现的时间早得远超任何人的想象，它们在地球吸积的数亿年间就已出现；人们原以为当时的地球是个

---

① 耶洛奈夫镇（Yellowknife），又译黄刀镇，加拿大西北地区的首府。

部分熔融、不适于居住的世界，看来这些发现也提出了与之相反的证据。

未来的超大陆

本章大部分篇幅都用来回望昔日的大陆圆舞曲了。但大陆仍在移动，那么在接下来的5000万年、1亿年甚或更远的未来，世界地图将会变成什么样？首先，合乎情理的假设是事物会沿着其当前的方向继续发展。大西洋会继续加宽，太平洋会收缩。曾经封闭特提斯洋的过程也会继续，阿尔卑斯山脉与喜马拉雅山脉之间的危险地带会发生更多的地震和山体抬升。澳大利亚会继续北移，赶上婆罗洲，继而转个圈撞上中国。在更远的未来，某些运动或许会反转。我们知道，大西洋的某个前身曾经开放过也封闭过，那么大西洋的洋壳最终会冷却、收缩，并开始再次下沉，或许会潜没到南北美洲的东岸下面，这些可能都无法避免。随后这些大陆会再度聚成一团。得克萨斯大学阿灵顿分校的克里斯托弗·斯科泰塞[①]预测，2.5亿年后会出现一个全新的超大陆——终极联合古陆，其间可能会有一个内陆海，那一切都将会是曾经不可一世的大西洋留下的遗物。

---

① 克里斯托弗·斯科泰塞（1953— ），美国地质学家，"古代地图计划"（Paleomap Project）的创建者，该计划的目标是绘制自数十亿年前开始的地球古代地图。

第六章
火山

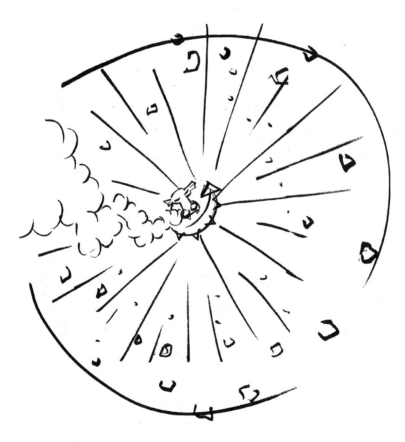

　　永无休止的构造板块运动、山脉的抬升与侵蚀,以及生物体的演化,这些过程都只有在地质学的"深时"跨度上才能够得

到充分认识。但某些正在我们这个星球上进行的过程可以在浓缩的刹那间把这一切在我们眼前铺陈开来,在便于人类观察的时间跨度上改变地貌、摧毁生命。那就是火山。把一个活火山地图叠加在世界地图上,可以清楚地看到构造板块的分界线,此二者的联系显而易见。例如,太平洋周围的火山带显然与板块边界相关。但流入这些火山的熔融态岩石是从哪儿来的?火山为何彼此不同:有些火山温和喷发,其熔融岩浆细水长流,而另一些的喷发则是毁灭性的爆炸?还有,某些火山远离任何明显的板块边界,例如夏威夷诸火山就位于太平洋中间,这又是怎么回事?

历史上不乏火山喷发目击者的记录和对这种现象的解释,其中有些是神话,有些是空想,但还有一些却惊人地准确。小普林尼[①]有关公元79年维苏威火山喷发的描述就是一份较为准确的记录。那场火山喷发毁灭了庞贝和赫库兰尼姆两座古城,他的舅舅老普林尼也不幸罹难。但在很长的历史时期中,人们并不了解火山喷发的原因,往往会认为火山喷发是火神或女神,如夏威夷女神佩蕾的杰作。在中世纪的欧洲,人们认为火山是地狱的烟囱。后来,有人提出地球是一颗逐渐冷却的星球,内部还有些残余的星星之火,通过裂隙的体系彼此相连。19世纪,我们如今所知的火山岩被普遍认为来自海洋的覆层,这就是所谓的水成论者,其理论与火成论者截然相反,后者认为这些岩石都是曾经熔融过的。在火成论者的理论发展和普及之后,很多人认为地球内部一定是熔融态的,这种观念直到地震学初露端倪之后才最终被抛弃。谜题之一是火山岩可以有不同的组分;有时甚至从同

---

① 小普林尼(61—约113),罗马帝国律师、作家和元老。在给塔西佗的信中,小普林尼描述了他的舅舅兼养父老普林尼丧生的那次维苏威火山爆发,也是在那次火山爆发中,庞贝城被毁。

一个火山喷发出来的火山岩也是如此。查尔斯·达尔文等人首先提出，密度高的矿物结晶析出，沉入岩浆，所以熔融物的组分会发生变化。达尔文对加拉帕戈斯群岛火山岩的观测结果为此说法提供了证据。至于达尔文提出的有关大陆漂移的想法，则直到20世纪中叶阿瑟·霍姆斯提出有关固态地球地幔中存在对流的想法时，才算是朝着科学的真相迈出了坚实的第一步。

岩石是如何熔融的

了解火山的关键在于了解岩石是如何熔融的。首先，岩石不必完全熔融，因而虽说熔融的岩石造就了熔融态的岩浆流体，大部分地幔仍保持固态。这意味着这种熔融物的组分不一定与大部分地幔的组分相同。只要表层岩的矿物颗粒相交的角度——所谓二面角——足够大，岩石就像多孔的海绵一样，熔融物也就会被挤出去。计算结果显示出它如何流动聚集，并以类似波的形式相当迅速地上升，在表面生成熔岩；一般来说，当熔岩达到一定数量，火山就会喷发。

熔融并不一定需要温度上升，它也有可能是压力下降的结果。因此，炙热的固态地幔物质组成的地幔柱在其上升过程中，所承受的压力下降，就会开始熔融。在地幔柱的例子中，这种情况在地下极深处即可发生。夏威夷喷发的玄武岩中氦同位素的比率表明，其生成于地下150公里左右。那里的地幔主要由富含矿物橄榄石的橄榄岩组成。与它相比，喷发而出的岩浆含有较少的镁和较多的铝。据估计，仅需4%的岩石熔融，便可产生夏威夷的玄武岩。

在洋中脊系统之下，熔融发生的位置要浅得多。这里几乎没有地幔岩石圈，炙热的软流层靠近表面。这里较低的压力可导致更大比例的岩石熔融，或达20%—25%，从而以适当的速率提供岩浆来维持海底扩张，并生成7公里厚的洋壳。大部分洋脊喷发都无人察觉，因为它们发生在2000多米深的水下，迅速熄灭，形成枕状熔岩。但地震学研究表明，在部分洋脊，特别是在太平洋和印度洋，洋底数公里之下存在岩浆房，不过也有证据表明大西洋中脊下也有岩浆房存在。在那里，地幔柱与洋脊系统叠合，冰岛即是一例；那里产生的岩浆更多，洋壳也更厚，乃至升出海面形成了冰岛。

夏威夷

夏威夷大岛不仅居民热情好客，那里的火山也一样十分友善。希洛镇附近有一座4000米高的冒纳罗亚（Mauna Loa）火山，但若说镇子背后这座火山有可能爆发，其威胁可能还不如远处的地震引发的海啸。希洛镇的北面和西面坐落着夏威夷群岛的其他岛屿和天皇海山链，描摹出太平洋板块横越其下地幔柱热点的漫漫征途。夏威夷大岛的南方是罗希海底山，这是夏威夷火山中最新的一个。迄今为止，它还没有露出太平洋水面，但已经在洋底建起了一座玄武岩高山，不久以后几乎一定会变成岛屿浮出水面。夏威夷的熔岩流动性很强，可以覆盖大片区域，而不太可能形成非常陡峭的斜坡。这样的火山有时也被称为盾状火山，可以大范围溢流玄武岩。某个特定的岩浆流往往会生成一个环绕岩流的隧道，外壳虽然凝固，内部仍然继续流淌着岩浆。岩浆断流之后，排干了的隧道便保持着中空状态。

最后一座停止喷发的大火山——冒纳凯阿火山（Mauna Kea）是地球上晴天最多的地方之一，也是一个国际天文观测台的所在地。我到那里游历的一天晚上，通过望远镜看到了位于冒纳罗亚火山侧面的普奥火山口的一次剧烈喷发。第二天，我乘坐直升机低空飞越了刚刚喷发出来的熔岩流。打开舱门，能够感受到炙热岩浆的辐射热，有好几处仍在流淌着熔岩，空气中有一丝硫磺的气味。但一切看来都很安全，甚至在火山口上方盘旋也是如此，不过要避开烟气与蒸汽的热柱。附近的基劳亚破火山口现今有过多次火山喷发，但那里竟然还有一个观测台和观景台。每隔几个星期，游客就能目睹一次新的喷发，起先往往能看到一片火幕，伴随着许多沿着裂缝排列的灼热岩浆喷泉。火中并无一物燃烧，但通过灼热奔流的熔岩释放的火山气导致白热的蒸汽喷射到数十米甚至数百米的空中。火山喷发可能只持续几个小时。尽管有高空喷射，但由于熔岩的高度流动性，火山的喷发并不是特别容易爆炸。附近火山观测台的火山学家们可以穿着热防护服接近熔岩，甚至火山口。他们很可能已经通过灵敏的测震仪网络以及测量岩浆上升流所带来的地心引力变化，探测到火山口的岩浆上升。有时，他们还能够从火山口直接采集未经污染的火山气样本并测量熔岩的温度。火山喷发的温度大约是1150℃。

普林尼式火山喷发

　　火山喷发的性质取决于岩浆的黏性及其所含的溶解气体和水的容量。在喷发早期，地下水全部会被飞速闪蒸成为水蒸气。随着压力的释放，气体从溶液中挥发出来，迅速扩张，有时甚至会爆炸。四下流淌的玄武岩中所含的气体适量，就能产生夏威夷的火喷涌现象。如果气体含量更大，会携带着细碎的火山灰和火山渣等固态物质。当岩浆仍含有大量气体时，初期的喷发会更加剧烈。如果它有时间在相对较窄的岩浆房沉积下来，表现就会温和得多。有时，气体和火山灰高高升入空中，导致火山灰

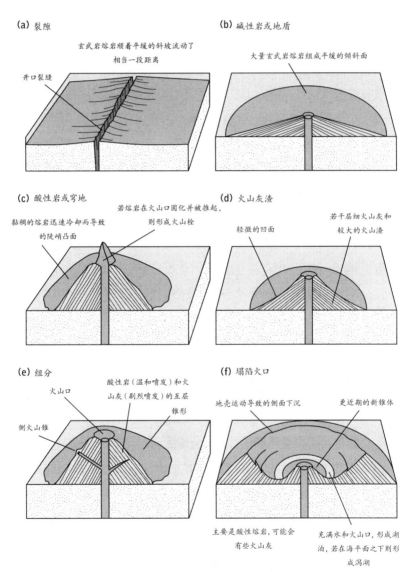

图17 以形状（不按比例尺）划分的主要火山类型

在空中大面积扩散。公元79年，人们目睹的维苏威火山喷发即是如此；这种现象被称为普林尼式火山喷发，得名于普林尼对其舅舅遇难情景的描述。

这样的喷发可以生成很多种不同的火山岩。火山灰和火山渣在落地之前质地坚硬，但在落地后则会形成松软的凝灰岩层。如果碎屑仍是熔融态的，则会形成熔结凝灰岩。靠近火山口的地方，会有大块岩浆被抛出。如果它们在落地时还是熔融状态，就会形成飞溅的炸弹，看出去有点像牛粪团。如果仍在飞行中的熔岩炸弹在空中凝结一层固态外壳，就会形成一枚剥层火山弹，看上去更像是一大块发酵面包。迅速猝熄的熔岩可能会形成一种被称作黑曜岩的火山玻璃。有时熔岩固化时内部仍有气泡，即所谓的气囊。有时熔岩中的气泡会形成泡沫，这样生成的浮岩密度很低，能够漂在水上。熔岩流的表面可能非常粗糙，像煤渣一样，在夏威夷人们称之为"啊啊"（ａａ）熔岩。（这是夏威夷语的一个单词，而不是人试图走过熔岩时发出的尖叫！）流体熔岩流形成的一层薄皮可能会起皱变成流线，生成绳状熔岩。有时会拉出细股的熔岩，这种效果有时被称为"佩蕾的发丝"。

太平洋火圈

只要避开火喷涌和快速流动的熔岩，夏威夷岛的火山喷发还是相当安全的。但大多数火山可不是这样。太平洋大部分都在火圈包围之中，那些火山的性情可要暴烈得多。当海洋板块下潜进大陆或岛弧的潜没带下，就形成了所谓的成层火山。这些火山常常是风景明信片的主角。日本富士山就是其中之一，它有着陡峭的圆锥形斜坡，积雪盖顶，火山口终年烟雾缭绕。但人们往往被这类火山的美丽外表所迷惑，对其险恶的行为视而不见。它们因频繁的地震和突然间剧烈的火山喷发而臭名昭著,比如日本的云仙岳火山和菲律宾的皮纳图博火山均在1991年显露出狰狞面目。它们之所以被称为成层火山,是因为熔岩、火山灰或火山渣交替的分层结构,喷涌所及的范围远大于山峰本身。

它们之所以比夏威夷火山表现暴戾,是因为岩浆不是来自地幔的清洁新生物质。其地下的下沉物质是旧洋壳,其中浸透了水,既有存在于气孔和裂隙中的液态水,也有与含水矿物结合的水。随着板块下潜,因为深度,可能也是摩擦力的作用,水被加热了。水的存在降低了熔融点,因而发生了部分熔融。压力非常大,以至于水在熔融物中很容易溶解,为其润滑,从而使得这部分岩浆向上挤压通过上面的陆壳。在它接近地表的过程中,压力下降,水分开始变成蒸汽逸出。这一过程非常迅速和剧烈,很像在充分摇动一瓶香槟之后拔去塞子,气体逸出的情形。

在上升过程中,岩浆会聚积在岩浆房中,直至积聚足够的压力喷发出去。在此期间,致密的矿物会固化并下落到岩浆房的底部。这些矿物,尤其是铁化合物,是令玄武岩变得黝黑致密的物质。留待喷发的熔融物颜色较浅,所含的氧化硅更多——在某些情况下含量高达70%—80%,而玄武岩中的氧化硅含量最多只有50%。它所形成的岩石被称为流纹岩和安山岩,是日本和安第斯山脉等地特有的岩石。喷发十分剧烈,不仅是由于含水量较高,还因为这种富含氧化硅的熔岩的黏性也高得多。它们不易流动,气泡不容易逸出。这种熔岩无法像夏威夷火山喷发那样形成喷射,只会一路轰鸣,喷薄而出。

圣海伦斯火山

近年来最著名的一次火山喷发就是一座这种类型的火山。

圣海伦斯火山位于美国西北的华盛顿州，那里是太平洋板块潜没之处。1980年初，那里还是一个松林湖泊环绕的美丽山脉，也是度假胜地。自1857年以来，它几乎没有过什么活动的迹象。然后，就在1980年3月20日，一系列小地颤累积成4.2级的地震，火山再度苏醒过来。地颤持续增加，引发了小型的山崩，直到3月27日，顶上火山口发生了一次大喷发，圣海伦斯火山开始喷涌出火山灰和蒸汽。盛行风将暗色的火山灰吹向一侧，另一侧则被白雪覆盖。火山呈现出黑白相映的景象。

截至那时，还没有熔岩喷发出来，从火山口逸出的只是蒸汽和被吹出的火山灰。但逸出的蒸汽是预警信号，表示火山下的炙热岩浆上升。地震活动继续，但测震仪也开始记录有节奏的连续地面震动，与地震的大幅震荡截然不同。这种所谓的谐波震颤据信是由火山下的岩浆上升而产生的。到5月中旬，所记录的地震达到一万次，在圣海伦斯火山北侧还出现了显著的隆起。地球物理学家向置放在隆起周围的反射器发射激光束，以此来测量隆起的速度，它以每天1.5米的惊人速度向北推进。到5月12日，隆起的某些部位比岩浆侵入开始前高了138米多。火山几乎被楔成两半，进入了极度不稳定的危险状态。

5月18日星期天一大早，基思和多萝西·斯托费尔乘坐一架小飞机越过火山上空时，突然注意到岩石和积雪向火山口内部滑落。几秒之内，顶部火山口的整个北侧开始移动。隆起部分在大山崩中塌陷下去。那情景就像拔出了香槟瓶口的塞子。里面的岩浆暴露在空气中。几乎瞬间便发生了爆炸。斯托费尔夫妇赶紧驾机俯冲以便提速逃生。美国地质勘探局的戴维·约翰斯顿就没那么走运了。在他使用无线电波从火山以北10公里处的观测站发送出最后一道激光束测量值之后一个半小时，北侧山坡塌陷下来，爆炸朝着他席卷

图18 1980年美国华盛顿州圣海伦斯火山喷发是近年来最壮观的景象之一,也留下了最精确完整的记录。火山灰和烟雾升腾近20公里进入大气层

而去。算他在内，共有57人在此次火山喷发中遇难。

爆炸开始的时间比山崩晚了几秒，但爆炸很快就占据上风。它散开的速度超过了每小时1000公里。在方圆12公里范围内，树木不但倒地，还被卷走。生灵涂炭，人造物荡然无存。30公里开外的树木也被掀倒，只在低洼处有零星的小片林子得以幸存。甚至在更远的野外，树叶也被热量烤焦，枝干惨遭折断。

横向爆炸过后不久，一道笔直的火山灰和蒸汽柱开始上升。不到10分钟就升到了20公里之高，并开始扩展成为典型的蘑菇云。盘旋的火山灰颗粒产生了静电，闪电造成了很多森林火灾。大风很快将火山灰吹到东面，航天卫星得以环绕地球跟踪其踪迹。火山灰落在美国西北部的大部分地区，厚度在1—10厘米之间。在9个小时的剧烈喷发期间，大约有5.4亿吨火山灰落在5.7万平方公里的土地上。

然而，这类火山喷发的另一个危险是所谓的火山碎屑流。它们是由爆炸散落的岩石或岩浆颗粒组成的，在一团高温气体中，以每小时数百公里的速度横扫开去。其温度和速度足以致命。1902年，西印度群岛马丁尼克岛上的培雷火山喷发，火山碎屑流扫荡了圣皮埃尔市，全市3万居民几乎全部罹难。颇有讽刺意味的是，两名幸存者中，有一位是由于被单独监禁在通风不佳的厚墙牢房里而活了下来。圣海伦斯火山的火山碎屑流并不比山崩碎片所到之处更远，然而在某些地方，喷发物质抵达古老的湖底，热量仍足以将湖水闪爆成蒸汽，那情形就像是次级火山喷发。公元79年，或许就是火山碎屑流摧毁了古城庞贝。

圣海伦斯火山本身的高度比喷发前矮了400米，中间产生了一个大大的新火山口。1988年又有数次爆发式喷发，1992年也有一次，但都不如第一次那样壮观。如今，山上到处都是科学仪器,它们完全能够捕捉到进一步活动的迹象。

过去的爆发

圣海伦斯火山的喷发或许看似可怕,但与史上和史前其他火山喷发相比,还算是小规模的。这次喷发将1.4立方公里的物质抛到空中。相比之下,1815年印度尼西亚的坦博拉火山喷发喷射出大约30立方公里的物质,而公元前5000年,美国俄勒冈州马札马火山的一次喷发产生了大约40立方公里的火山灰。1883年,喀拉喀托岛(位于爪哇岛西面,而非电影①片名中所说的东面)喷发,在洋底留下了一个290米深的火山口。共有3.6万人伤亡,大多数溺死于随后的海啸,在海啸中,40米高的巨浪将一艘蒸汽船深深搁浅在丛林中。大约公元前1627年,轮到爱琴海的桑托林岛——或称希拉岛——喷发。那次喷发发生于弥诺斯文明②青铜时代的鼎盛时期,很可能成为这一文明衰败的原因之一,也是关于失落的亚特兰蒂斯大陆的传说的来源。在地质学的时间尺度上,这些也不过是一系列剧烈喷发中距今最近的几次而已。

火山的剖析

人们关于火山的刻板印象,无非是圆锥形的山峦坐落在水池一样的岩浆房上,从顶上火山口中喷出岩浆,然而事实上,很少有真实的火山与这种刻板形象相符。西西里岛的埃特纳火山是学界研究最全面的火山之一,它显然要复杂得多。这是一座非

---

① 指1969年的美国电影《喀拉喀托:爪哇之东》,曾获得1970年美国奥斯卡最佳视觉效果奖提名。

② 主要集中在爱琴海地区克里特岛的古代文明,其持续时间大约是公元前2700—前1450年。

常活跃的火山，地质年龄很可能只有大约25万年。当然它不可能一直持续着近30年来人们所观测到的活跃程度，否则它应该更大一些。它与维苏威火山以及它北面的武尔卡诺岛和斯特龙博利岛等火山岛不同。那些是由爱奥尼亚海海底的潜没作用所提供的喷发物质组成的成层火山。与之相反，埃特纳火山可能源自地幔柱。但它似乎天性善变。对不同年代的熔岩组成的测量显示，近来它开始呈现出更多前一种火山的特性，也就是说，更像是它北方那些由潜没作用注入物质的火山了；此外，其喷发的性质看来也的确在改变，变得越来越剧烈，具有潜在的危险。

在埃特纳这样的火山下面进行测量和研究会非常复杂。并不存在什么现成的中空管道系统坐等岩浆的到来；岩浆必须经由阻力最小的路线强行上升。在地幔柱中，首选路线很可能是易于被泪滴形的上升岩浆块推开的低密度物质柱。在更坚硬的地壳中，岩浆必须找到一条穿过裂隙的路线。大型火山非常重，会使其所在的地壳超载，形成同心裂纹网。火山活跃期停止后，它会沿着这些裂纹塌陷，生成一个宽阔的破火山口。随着岩浆继续上升，它会强行进入裂纹，形成同心的圆锥形薄板或环状岩墙的脉群。火山内部的岩浆上升会造成其隆起，并爆裂成一系列小型的地颤。在埃特纳火山这个例子中，顶上火山口显示出近乎持续的活跃度。我曾在静止期爬上陡峭而疏松的火山渣锥向内窥视。即使在那时，地面摸上去仍是温热的，空气中也有硫磺的气味。火山口喷发出水蒸气，发出的声音与我想象中巨人或恶龙的鼾声没有什么不同。

有时，恶龙醒来，火山口边缘便不再是安全的立脚之所。喷发开始时，直径达一米之大的炙热岩石块可能会被抛到空中。1979年的一次火山喷发便是如此开始的，但随后被一场大雨浇

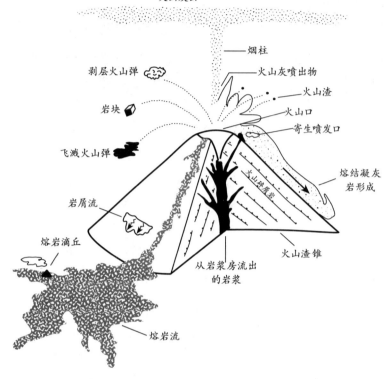

图19 喷发中的复合火山——如埃特纳火山——的一些主要特征

得陷入沉默,导致了火山口内侧的滑塌。然而,压力累积起来并引发了爆炸。不幸的是,当时火山口周围还站着很多游客。那场喷发中有30人受伤,9人遇难。英国公开大学的约翰·默里博士回忆起1986年的另一次火山喷发,当时他在观察一场貌似相当普通的喷发,整个下午火山活动缓慢。有火山弹落在火山口外200米范围内,地质学家们还很安全。随后,火山弹的落地范围突然扩大到逾2公里。大块的岩石从地质学家的头上呼啸而过,落在他们身旁。在统计学意义上,被岩石击中的概率并不大,但约翰·默里说身临其境的感觉可没这么淡定。

约翰·默里及其同事多年来一直在观测埃特纳火山,他们越来越熟悉喷发来临前的征兆。他们会使用测量技术和全球定位系统来定位山腹因其下的岩浆上升而发生的轻微隆起,还会监测地表裂缝被迫张开而引发的地颤。重力测量可以显示岩浆上升时的密度。喷发平静下来之后,测量者还会监测山坡,他们特别关注的是卡塔尼亚城①上方的东南侧陡峭山坡。1980年代初期,该山坡有些地段在一年之内下沉了1.4米,有人担心山坡会就此塌陷,甚至会降低内部岩浆所受的压力,从而引发像圣海伦斯火山喷发时那样的横向爆炸。或许早在公元前1500年左右就已经发生过这种情况,迫使古希腊人放弃了西西里岛东部。

火山喷发一般从顶上火山口开始,但是,一旦初期的气体释放,岩浆便会强行穿过山腰的裂隙,有时会危及附近的村落。人们曾试图通过开凿另外的通道和铲平土方来让岩浆流转向,或者用水淋浇甚或爆破来阻止岩浆流。这些方法有时可以挽救家园,有时却无济于事。1983年,一道岩浆一直流到接近地质学家居住的萨皮恩札饭店外才停了下来。时至今日,与山峦的威力相比,人类压制火山力量的尝试仍然显得微不足道。

火山与人类

火山灰和参差不齐的岩浆流会以令人惊讶的速度迅速分解,产生富饶肥沃的土壤。健忘的人们总是聚集在火山周围,在那里建起农场、村落乃至城市。如今,监测火山,并在其内岩浆上升、喷发即将来临时至少获得某些警报都是可能的。但即使在那样的时刻,劝说人们离开有时也绝非易事。在维苏威火山下的

---

① 意大利南部西西里岛东岸一城市,也是卡塔尼亚省的首府。历史上以地震频繁闻名。

主要的火山

名称	海拔（米）	大喷发时间	最后一次
苏联别济米安纳	2800	1955—1956	1984
墨西哥埃尔奇琼	1349	1982	1982
南极洲埃里伯斯	4023	1947, 1972	1986
意大利埃特纳	3236	频繁	2002
日本富士山	3776	1707	1707
爪哇岛加隆贡	2180	1822, 1918	1982
冰岛赫克勒	1491	1693, 1845, 1947—1948, 1970	1981
冰岛海尔加费德	215	1973	1973
阿拉斯加州卡特迈	2298	1912, 1920, 1921	1931
夏威夷基拉韦厄	1247	频繁	1991
苏联克柳切夫斯科耶	4850	1700—1966, 1984	1985
苏门答腊岛喀拉喀托	818	频繁，尤其是1883	1980
圣文森特岛苏弗里耶尔	1232	1718, 1812, 1902, 1971—1972	1979
美国拉森峰	3186	1914—1915	1921
夏威夷冒纳罗亚	4172	频繁	1984
菲律宾马荣	2462	1616, 1766, 1814, 1897, 1914	2001
蒙特塞拉特		1995年前一直休眠	1995—1998
扎伊尔尼亚穆拉吉拉	3056	1921—1938, 1971, 1980	1984
墨西哥帕里库廷	3188	1943—1952	1952
马丁尼克岛培雷	1397	1902, 1929—1932	1932
菲律宾皮纳图博	1462	1391, 1991	1991
墨西哥波波卡特佩特	5483	1920	1943
美国瑞尼尔山	4392	公元前1世纪, 1820	1882
新西兰鲁阿佩胡	2796	1945, 1953, 1969, 1975	1986
美国圣海伦斯	2549	频繁，尤其是1980	1987
希腊桑托林岛	1315	频繁，尤其是公元前1470	1950
意大利斯特龙博利	931	频繁	2002
冰岛叙尔特塞	174	1963—1967	1967
日本云仙岳	1360	1360, 1791	1991
意大利维苏威	1289	频繁，尤其是公元79	1944

那不勒斯湾这样人口密集的地方，及时疏散人群既不现实也不可能。在南美洲和其他地方那些不甚著名的火山，至今甚至没有学者对其作过任何冲击力方面的学术研究。

移动缓慢的岩浆流和更加致命的火山碎屑流并非仅有的危险。在喷发的火山周围聚集起来的尘埃和水蒸气的喷发云会造成大雨，若与火山上融化的雪水配合，会导致灾难性的泥石流或火山泥流。1985年，哥伦比亚的内瓦多·德·鲁伊斯火山就曾发生过这样的情况，泥石流沿山坡一路扫荡，造成大约22 000人死亡。威胁甚至可能毫无行迹可觅。喀麦隆境内尼奥斯湖的深水中累积溶解了大量的火山二氧化碳。1986年的一个寒夜，湖面上致密冰冷的水体突然下沉，把富含气体的水带到湖面并释放了其中的压力。这就像打开一瓶充分摇动的苏打水一样，密度大于空气的气体突然释放席卷了山谷，导致附近村里的1700人在梦乡中窒息而死。

火山的力量或许无法阻挡，但通过审慎规划和仔细监测，人类可以学着相对安全地与之共存。

第七章
地动山摇之时

全速横穿大洋的超级油轮有着很大的动能,制动距离长达数十公里,然而没有什么能够阻挡整块大陆。我们已经听过印度和亚洲历时5500万年的慢动作碰撞的故事,其他构造板块也都在发生相对位移。板块的相互摩擦引发了地震。大地震地图所显示的板块分界线甚至比火山地图还要清晰。

全球定位系统测量显示了构造板块如何以每年几厘米的速度缓慢而稳定地滑过彼此。但在接近板块边缘处,滑动就没那么流畅了。在有些地方,运动仍然稳定,没有发生大地震,岩石仿佛被涂抹了润滑剂,抑或变得非常柔软,以至于它们的移动简直可以称之为蠕动。但很多板块边界却动弹不得。然而大陆持续移动,张力不断累积,最终岩石不堪忍受而突然破碎,地震就发生了。

随着洋壳潜没进地幔,也有些地震发生在很深的位置。但大多数地震都发生在顶部15—20公里处,那里的地壳又热又脆。岩石沿着所谓的断层线破碎,传出地震波。地震波看来是从地下的震源沿断层散发出去的。震源上方地表上的那个点叫作震中。

地震的震级

里克特和梅尔卡利震级表为地震划分了等级。前一种测量的是实际的波能,后一种则标记了破坏的效果。里克特震级表是用对数表示的,因此只使用数字1—10便可标记所有地震:从地震活跃区几乎无法察觉的频繁日常颤动,到有据可查的最大地震——迄今为止,最大规模的地震发生于1960年的智利海岸,其在震级表上的测量值为9.5级。震级表上每个点之间的能量差别是30倍。因此,例如,7级地震很可能比6级的破坏力大得多。令人啼笑皆非的是,查尔斯·里克特这位为震级表命名的加州地震学家1985年去世之后,他的许多个人记录却在1994年洛杉矶附近北岭镇的一场6.6级地震所引发的火灾中毁于一旦。

世上最著名的裂缝

在美国加州,地震几乎就是家常便饭。巨大的太平洋板块在持续运动中,它没有下潜到美洲大陆之下,而是在所谓平移断层与大陆擦身而过。接合位置几乎从来都不是一条直线,因此,主要断层线上的扭结导致很多平行和交叉的裂缝或断层。其中

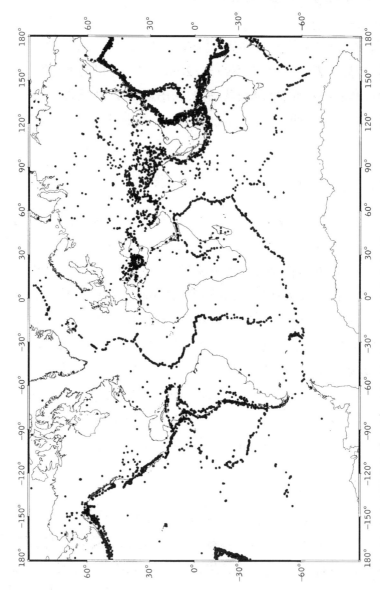

图20 过去30年来地壳大地震（5级和以上）的分布。大多数聚集在构造板块的边界上，但也有少数发生在大陆腹地

大多数位置经历了频繁的小型地震,任何一处均有可能成为大地震的中心。包括板块边界本身在内,最著名的裂缝要数圣安德烈亚斯断层了。它起自加州南部,在洛杉矶内陆曲线行进,一路向西北直达旧金山,通向大海。它在1906年恶名加身,当时旧金山毁于一场大地震,随后可怕的火灾烧毁了所有的木质房屋。

洛杉矶与旧金山之间是一片不毛之地,很容易在光秃秃的山峦中辨认断层的痕迹。有时,斜坡上的一个微小变化即可标记其行踪。有时,可以看见它切断了地形,仿佛有一只大手持刀划过地图。断层似乎直线行进了100英里。我曾在洛杉矶和旧金山中间的一条高低不平的机耕道上沿着这条断层行进。断层东面是坦布洛山脉低矮的侵蚀丘陵,西面则是干燥的卡里佐平原缓坡,向下直通圣路易斯-奥比斯保和太平洋。从丘陵地带下来是一些干涸的河床。在河床与斜坡基部之处,似乎发生过一些奇怪的事情。那些河流并没有直接流向西方,而是向右急转90°,沿着丘陵基部向北走了几十米,然后又向左急转弯继续流向大海。华莱士溪是其中规模最大的河流之一,这条得名于美国地质勘探局的罗伯特·华莱士的河流深深地嵌入柔缓的丘陵斜坡。它在横穿断层时错位了130米。它起初必然是直接流下山坡的,沿途切出深深的河道。在一系列地震中,平原西部突然向北倾斜,河床自然也跟着转向了。冬汛无法在高耸的河岸上切出一条新的河道,所以这些溪流沿着断层流淌,直到再次与原先的河床会合。这不是一蹴而就的。罗伯特·华莱士及其同事综合使用了挖方和放射性碳测定年代技术,计算出了地质时期。史书中唯一一个记录发生在1857年,它解释了最后9.5米的平移。距离这次平移最近的两次移动都发生在史前,分别将河道平移了12.5米和11米。平均下来,圣安德烈亚斯断层在过去13 000年以每年34毫米

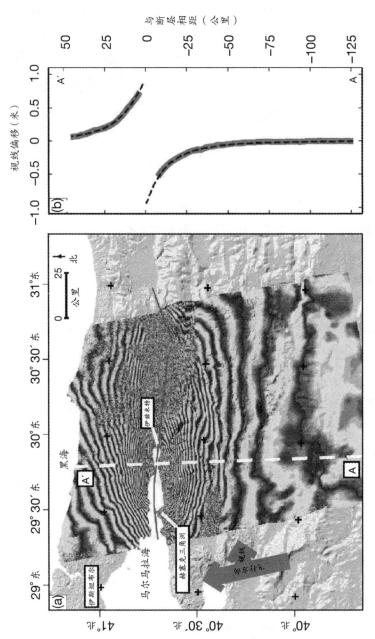

图21 1999年土耳其伊兹米特地震前后数据相结合之后的卫星雷达干涉图，显示出地面的运动

的速度滑动。如果保持这一速度，那么2000万年后，洛杉矶将会挪移到旧金山的北方，两个城市间的距离与当前一样，只不过现在它是在旧金山的南面。当然了，加州人都知道这条道路并不平缓，他们为此可算是吃尽了苦头。

测量移动距离

在大陆尺度上测量以米或厘米为单位的移动距离，在过去几乎是不可能的，但如今就相对容易了。像美国加州和日本这样的断裂带上被放置了各种仪器。尤其是与全球定位系统连接的接收器，可以保证连续监视这些断裂带在地表上的位置。如果它们联入自动监测网，就可以立即向有关方面通知地震发生的准确位置和严重程度。正如我们将要看到的，它们还有助于地震预警。太空可以传送更清晰的实况图像。配备了合成孔径雷达的遥感卫星可以记录地面的形状，精确度非常高，以至于将地震前后分别拍摄的两张图片叠加后，可以生成干涉图样，显示出移动断层的精确区段及其运动。

板块中部的地震

就算表面上看来十分坚硬的大陆板块也受到压力和张力，它们有时也会移动。在美国简短的历史记录中，最大的地震并不是发生在加州，而是在美国东部。1811年，圣路易斯市附近的边城新马德里被3场里克特震级高达8.5级的大地震所撼动。这3场地震威力强大，足以摇动波士顿教堂里的吊钟，如果当时广阔的密西西比平原上存在大型的现代城市，也定会被夷为平地。迄今也不确定那场地震是由于在密西西比河沉积物的重压之下大地下沉所导致的，还是威力无穷的密西西比河本身恰恰就是地壳延展的产物。原因可能在于这里是某个海洋企图开放的备选线路，尽管最终它选择了阿巴拉契亚山脉另一侧的大西洋。但也许它是在作另一番尝试。无论原因如何，如果今天在新马德里再来一次地震，所造成的破坏将是不可估量的。

深层地震之谜

通过绘制地震深度图，我们就有可能跟踪海洋岩石圈在潜没带下降的位置，如太平洋板块潜入南美洲安第斯山脉之下的具体位置。在起初的200公里左右，岩石冷脆，因而会破碎，又因为接近表面，所以就产生了地震。但某些地震的震源似乎要深得多，有的深达600公里，那里的热量和压力会让岩石变得软韧，使它们更易变形而非破碎。一个可能的解释是，这些深层地震或许是由于整层结晶正在经历着相变，从海洋岩石圈中的橄榄石结构变为地幔中密度更大的尖晶石型结构。反对这一学说的论据是，这一过程只会发生一次，而同一地点迄今已经发生了好几次有记录的地震。但这也可能是因为有连续若干层橄榄石在经历相变。

听天由命

2001年1月，整个印度西北部地动山摇，这场毁灭性的地震震中位于古吉拉特邦普杰市。此乃印度与亚洲的洲际碰撞余波未了。印度与中国西藏之间的相对运动仍在以每个世纪大约2米的速度持续累积。20世纪已经发生了一些严重的喜马拉雅地震，但一定有很多地区累积了大得多的张力。2米的滑坡便有可能产生7.8级的地震。但在印度从下方推入喜马拉雅山脉的逆断层中，某些部分已经积累了相当于4米滑坡的张力。实际上，某些地

图22 近年来城市地震中的高速公路隆起导致了惊人的人员伤亡。这是1995年日本神户市的景象

区500多年都没发生过严重的地震了。这样的"大地震"的确会是毁灭性的。尽管上个世纪以来建筑标准有所提高,普杰地震的证据表明,如今一次震级相当的地震可能造成的死亡人口比例与100年前无甚差别。而与此同时,处于危险之中的人口数量增加了10倍甚至更多。如果1905年的坎格拉①地震今日重现,遇难人数很可能会达到20万。如果恒河平原的某座大城市发生了地震,这个数字可能会上升一个量级。东京,另一个人口密集的地震带,从1923年至今尚未发生过大地震。如果那里现在发生一场大地震,就算日本的建筑标准有所提高,估计也会造成7万亿美元的破坏,这也许会导致全球经济的崩溃。

为地震而设计

人们常说,取人性命的是建筑物,而非地震。当然,地震中的大多数遇难者都死于倒塌的建筑物和后续的火灾。建筑物是否会在地震中倒塌,受到很多因素的影响。地震的力量显然十分重要,但震动持续的时间也很重要,其后便是建筑的设计。就小规模的结构而言,柔韧材料比脆硬的材料好。就像树木可以在风中摇曳,木框架的建筑物也可以在地震中摇摆而不致倒塌。重量轻的结构在倒塌之时不易致人丧命。但日本住宅传统使用的木材和纸张隔断墙更容易着火。地震中最糟糕的建筑大概就是砖石建筑以及加固质量很差的混凝土框架了;这在较为贫穷的地震易发国家极为常见。1988年亚美尼亚的斯皮塔克市附近的地震和1989年美国旧金山附近的洛马普列塔地震都是7级,但前者导致10万人死亡,而后者的死亡人数只有62人,这很大程

---

① 印度北部喜马偕尔邦人口最稠密的地区,位于喜马拉雅山脉的山脚下。

度上是因为加州有非常严格的建筑物规范。那里的高层建筑非常坚固,并且不会与地震波的频率产生共振。许多高层建筑在地基内都安放了橡胶块来吸收震动。在日本,一些摩天大楼在屋顶处配有重物系统,可以快速移动以抵消地震引起的晃动。

地面液化之时

如果你曾经站在湿润的沙滩上上下晃脚,你或许会注意到,水会在沙中上升,你的脚会陷进去;沙子液化了。地震摇动潮湿

图23 加州的圣安德烈亚斯断层并非地壳上的一条单一裂缝。连这张地图也是经过简化的

的沉积物时,也会发生同样的现象。1985年墨西哥城地震的震中远在400公里之外,但城里还是有很多建筑被毁。那些建筑建在一个古湖的填拓地①上,地震波在淤泥中往复共振了将近3分钟,导致地面液化,便无法再支撑其上的建筑。无论建筑物的地基有多深,一旦地面液化,都无法再提供多少支撑了。在1906年和1989年旧金山附近的两次地震中,毁坏最严重的建筑物都位于海港区,就是建在填拓地上的。

火灾

地震袭击城市时,最大的危险之一便是火灾。在1906年的旧金山地震和1923年的东京地震中,死于火灾的人数都多过死于地震本身的人数。炉灶倾倒就很容易引起火灾,在木质建筑物的杂乱废墟中快速蔓延,破裂的煤气管道更是为之添柴加油。旧金山的消防力量不足;消防车都被困在车库里或堵塞的街道上,破裂的给水主管将城市的给水系统消耗殆尽。如今,像旧金山这样的地震易发城市都在开发燃气和自来水的所谓"智能管道"系统,在压力由于管道破裂而突然降低的情况下,该系统可以迅速地自动关闭相应区段。

拯救生命

地震期间最安全的地方当属开放、平坦的乡间,而恐慌则是最要不得的。在城市里,室外随处会有下落的玻璃和砖石,还不如待在室内的楼梯等坚固结构之下来得安全。日本和美国加州的学童都接受了如何进行自我保护的常规训练。然而,在真正

---

① 由原有的海域、湖区或河岸填埋形成的陆地。

的地震发生时，大多数人还是会愣在原地或惊慌失措地跑向室外。如果人们被困在倒塌的建筑物中，搜救人员可以使用一整套热敏相机和探听装置来找到他们，悲剧固然不可避免，但每一次灾难过后，都会发生不可思议的营救奇迹。

机会与混乱

在某种程度上，地震是能够有把握地预测的。旧金山、东京和墨西哥城等城市必将经历下一次地震。但知道这一点对于住在那里的人没什么用处。他们想准确知道的是"大限"何时将至，地震又有多严重。但这些正是地质学家无法预测的。就像天气一样，地球也是一个复杂的系统，微小的起因就可以导致巨大的结果。亚马孙河畔那只虚构的蝴蝶振动一下翅膀就能影响欧洲的天气，同样，卡在断层中的小卵石也可以引发地震。人类大概永远不可能完全准确地预测地震，但以概率预测地震的效果越来越好了；距离地震发生的时间越近，预测的准确率越高。

传统的地震前兆

在科学仪器出现以前很久，人类就一直在寻找地震预警，向他们通报即将来临的地震。特别是中国人，他们相当擅长观察奇怪的动物行为、水位的突然变化和水井中的含气量，以及其他地震前的征兆。正是利用这些指标，在1975年的一场毁灭性地震发生前数小时，海城市便被疏散一空，拯救了数十万人的生命。但一年后，24万人死于唐山，事先没有任何预警。其他线索包括微弱的光电闪烁，可能是由于矿物结晶受到挤压而产生的，就像按下压电式气体打火机会产生火星。有人对动物感知地震迫近的方式进行了认真研究；例如，日本有人观察鲶鱼是否会因为电子干扰而表现异常。但鲶鱼的异常行为有哪些？何况有多少户人家会监测鲶鱼呢？另有证据表明，在大地震来临之前会有极低频率的电磁波。不过这么说到底，最准确的指标看来要属穿行过地面的地震波了。

把握先机

多数大型地震发生之前都有前震。问题在于很难说清某个小型的地颤是一个独立事件，还是大地震的前奏。但前震可以改变概率。根据历史记录，我们可以说，在接下来的100年内的某个时间，有可能发生一场大地震。但这样一来，明天发生地震的概率就变成了1/36 500。每年可能会有10次小型地颤，其中任何一个都有可能是大地震的前震。因此，探测出小型地颤也就把未来24小时发生地震的概率提高到1/1000。了解到断层的具体位置、它们上次开裂是什么时候，并在所有适当的位置安放仪器，有时可将预测的准确率提高到1/20。但这仍然相当于说明天有95%的概率**不会**发生地震，基本上不能在电台广播上播报这一统计数值并以此为由疏散市民。不过它足以通知紧急部门待命，并停止运输危险化学品。

实时警报

预测地震也许总是很难，但地震发生时是可以切实探测到的。这可以变成一种预警系统。1989年美国加州的洛马普列塔地震之后，这一系统得到了验证。尼米兹高速公路的隆起路段已经部分坍塌，救援人员试图救出困在下面的驾车人士。巨大的混凝土板很不稳定，余震会让它们轰然倒塌。地震的震中远在近100公里之外，因此，设置在断层中的传感器能够利用无线

大地震

位置	年份	震级	死亡人数
印度普杰	2001	7.7	20 085
萨尔瓦多	2001	7.7	844
秘鲁	2001	8.4	75
中国台湾	1999	7.7	2400
土耳其	1999	7.6	17 118
阿富汗	1998	6.1	4000
伊朗北部	1997	7.1	1560
俄罗斯（库页岛）	1995	7.5	2000
日本（神户）	1995	7.2	6310
加州南部	1994	6.8	60
印度南部（奥斯马纳巴德）	1993	6.4	9748
菲律宾	1990	7.7	1653
伊朗西北部	1990	7.5	36 000
旧金山（洛马普列塔）	1989	7.1	62
亚美尼亚	1988	7.0	100 000
墨西哥城	1985	8.1	7200
也门北部	1982	6.0	2800
意大利南部	1980	7.2	4500
伊朗东北部	1978	7.7	25 000
中国唐山	1976	8.2	242 000
危地马拉市	1976	7.5	22 778
秘鲁	1970	7.7	66 000
伊朗东北部	1968	7.4	11 600
中国古浪南山	1927	8.3	200 000
日本	1923	8.3	143 000
中国甘肃	1920	8.6	180 000
意大利墨西拿	1908	7.5	120 000
旧金山	1906	8.3	500
印度加尔各答	1737	—	300 000
日本北海道	1730	—	137 000
中国陕西	1556	—	830 000
土耳其安条克	526	—	250 000

电以光速发送警报，救援人员会在以声速行进的地震波到达之前25秒收到预警，让人们有时间清场。未来，可以利用这一系统给出地震主震的简短警报。例如，从洛杉矶外的主断层线发出的冲击波需要一分钟才能到达该市。无线电警报或许不足以疏散人群，但联入计算机系统后，警报有助于银行保存账目、电梯停止运行并开门、自动阀封锁管道、急救车驶离建筑物。

结语

要介绍一个美妙的星球,本书的确只是个短小的读本。我努力概述了某些关键过程的运作,它们分别发生在我们头顶的星空和脚下的大地。我试图解释这些充满生机的过程如何在地表彼此互动,才让我们拥有了这个因为了解所以深爱的世界。那些过程产生了丰富多样的地形、岩石和生命群体。我无意介绍是怎样美丽的矿物和晶体组成了这个星球上的岩石;也没有探讨某些过程的细节,在这些过程中,岩石因构造力而上下颠簸,被风雨冰雪雕琢成我们生活的地球上这令人屏息赞叹的壮丽地形。我没有深究岩石的地下残余如何淤积在沉积层中,也没有探察它们如何产生了肥沃土壤,让我们的食物链得以存续。我更没有细说这个星球最最妙不可言的产物——生命,以及这个世界的物理力量如何与化学作用和自然选择协力,让我们的星球生机盎然。所有这些都该有各自的专著,一一进行更加完整的介绍。

然而,我的确相信我们的星球非常特别,如果不是各种地球物理过程的罕见组合,至少就我们所知,生命不会有机会如此辉煌地绽放。我试图在本书中展现世间万物是如何彼此依赖的。没有水,岩石便不会得到润滑,花岗岩不会形成,我们也不会拥有大陆上的大块陆地。没有水,也就没有云和雨;只有风卷黄沙的沙漠地形,是不大可能孕育生命的。没有液态水,生命的化学过程就无法起效，我们所知的生命也不会出现在地球上。没有生命，也就不会有对大气成分的回馈机制，至少到目前为止，正是这一机制让气候尚可承受。没有生命，现在的地球要么是个雪球世界，要么是个超热的温室。

虽说在最近这十亿年来的大部分时间里，环境适宜、万物生长，然而时至今日，我们的命运仍要听凭地球的摆布。火山和地震等构造力量比水灾、干旱和风暴等大气力量更为凶猛。它们吞噬了无数生命，让数百万人生活无以为继。然而无论如何，我们幸存下来了。在很大程度上，我们像蚂蚁一样在地表忙忙碌碌，对宏阔的图景不知不觉。但即使如此，人类本身也已经成为塑造星球的一股强大势力。城市化和农业、土木工程和污染，这些人为过程改变了大片地表的样貌。我们也为此付出了代价。当前动植物物种灭绝的速度甚至比白垩纪和二叠纪末期的物种灭绝速度还要快。如今，大气成分的变化及其所导致的气候变化看来比最后一次冰川期以来的任何时候都要迅速，持续的时间可能也要长得多。

我们早已不再是这个星球的受害者，而变成了它的托管人。我们却恩将仇报，对土地粗暴轻率地贪婪，对污染置若罔闻地轻忽。这样做是要承担风险的。我们仍然别无退路，毕竟所有的人都住在同一个星球上。我们应该照顾好这个星球，为它承担起责任。但也要开启征程去寻找新的家园，让新技术带领我们飞往外星。

无论我们怎样小心呵护，这个世界都不会永远存在下去。地球随时都会因小行星或彗星的撞击而毁灭，或者被附近某颗爆炸恒星的辐射穿在炙叉上烘烤。我们还有可能因为全面核战而更早地面临大致相同的结果。最终，在大约50亿年的时间里，太阳会耗尽其核心的氢燃料并开始扩张成一个火红的巨星。最新的估计认为，白热的气体倒不会抵达地球这么远的地方，不过它必然会吞噬水星和金星。它会把我们美丽的世界烧成灰烬，烧干海洋和大气层，让地球不再宜居。但就算对一个星球来说，50亿年也是一段漫长的时间。所有的物种都会灭绝，从统计学角度来说，人类根本不可能存活500万年，更不用说50亿年了。或许那时会有一种新的生命形式统治地球，或许我们会演化或自我设计成不同的物种，又或许我们的后代会找到某种方式把记忆和意识封存进永生的机器。总而言之，我是个乐观主义者，喜欢想象未来的行星科学家探索和开拓新世界，并把它们拿来与被我们称为地球的这个生气勃勃的行星相媲美。

索 引

（条目后的数字为原文页码）

A

Acasta 阿卡斯塔 96

accretion 吸积 33

Alexander Terrane 亚历山大岩层 85

Alps 阿尔卑斯山脉 91

Alvarez, Walter and Louis 沃尔特和路易斯·阿尔瓦雷茨 29

Amazon 亚马孙河 57

andesite 安山岩 105

anhydrite 硬石膏 63

anoxic events 缺氧事件 62

Armenia 亚美尼亚 124

asteroid impact 小行星撞击 29

asthenosphere 软流层 17, 100

Atlantic conveyor belt 大西洋传送带 12

atmosphere 大气层 9

atmosphere, circulation of 大气环流 12

atmospheric composition 大气成分 6

aurora 极光 8

Australia 澳大利亚 96

B

bacteria 细菌 64, 69

basalt 玄武岩 4, 30, 36, 68

Bermuda Triangle 百慕大三角 64

Bhuj 普杰 122

black smokers 黑水喷口 69

Broecker, Wally 沃利·布勒克尔 12

building regulations 建筑规范 124

C

California 加利福尼亚州 118

carbon 碳 14, 26

carbon compensation depth 碳补偿深度 62

carbon cycle 碳循环 16

carbon dioxide 二氧化碳 14

chalk 白垩 62

Challenger HMS 英国皇家海军"挑战者"号 55

chemical weathering 化学风化作用 18, 56, 91

climate 气候 10, 63

climate change 气候变化 14

coal 煤 95

coccolithophores 石灰质鞭毛虫 62

comets 彗星 56

continental crust 陆壳 77

continental reflection profiling 大陆反射剖面 93

continental shelf 大陆架 57

continents of the future 未来的大陆

134

continent, first 第一块大陆 95, 97
convection 对流 18, 45
copepods 桡足动物 17
core 核 17, 33, 47
core rotation 核旋转 52
core, inner 内核 48
Coriolis effect 科里奥利效应 12, 50
Courtillot, Vincent 文森特·库尔蒂耶 87
Cox, Keith 基思·考克斯 87
cratons 克拉通 92
creep 蠕动 37, 116
Cretaceous period 白垩纪 29, 87
cyanobacteria 蓝藻细菌 32

D

D" or D double prime layer 核幔边界（D"分界层）39, 44
Daisyworld 雏菊世界 15
Dartmoor 达特穆尔高地 94
Darwin, Charles 查尔斯·达尔文 24, 99
day length 昼长 44
Deccan Traps 德干地盾 87
deep earthquakes 深层地震 41, 121
diamond 金刚石 42, 137
diamond anvil 金刚石砧 39, 52
drilling 钻探 36

Dryas 仙女木属 13
du Toit, Alex 亚历克斯·杜托伊特 78

E

Earth 地球 33
Earth systems 地球系统 3
Earth, formation of 地球的形成 33
earthquake early-warning systems 地震预警系统 129
earthquake prediction 地震预报 126
earthquakes 地震 115–129
epicentre 震中 116
Etna 埃特纳火山 110
evaporite 蒸发岩 63
extinctions 灭绝 29, 30

F

Fiji 斐济 42
fire 火灾 125
flood, biblical 《圣经》记载的大洪水 22
flysch 复理石 91
folds 折叠 80
foreshocks 前震 127

G

gabbro 辉长岩 69, 74

135

Gaia 盖亚 15

gas hydrate 天然气水合物 *See* methane hydrate geological column 见甲烷水合物地质柱状剖面 26

Gibraltar 直布罗陀 63

Global Positioning System 全球定位系统 83

global warming 全球气候变暖 14

Gondwanaland 冈瓦纳古陆 85

Grand Canyon 大峡谷 26

granite 花岗岩 4, 92, 94

gravity 地心引力 46

Great Rift Valley 大裂谷 75

greenhouse effect 温室效应 13

greenhouse gases 温室气体 11

Greenland 格陵兰 31

Gulf of Mexico 墨西哥湾 64

Gulf Stream 湾流 12

H

Hatton Bank 哈顿滩 71

Hawaii 夏威夷 101

helium 氦 36

Hess, Harry 哈里·赫斯 78

Himalayas 喜马拉雅山脉 74, 89

Holmes, Arthur 阿瑟·霍姆斯 26, 65, 78, 100

Hoyle, Sir Fred 弗雷德·霍伊尔爵士 1

human ancestors 人类的祖先 21, 91

Hutton, James 詹姆斯·赫顿 18, 23, 78

hydrothermal vents 热液喷溢口 32, 69

hypocentre 震源 116

I

Iapetus Ocean 古大西洋 74

ice age 冰川期 11, 13, 60

ice cores 冰核 15

Iceland 冰岛 71, 101

igneous rocks 火成岩 4

India 印度 86

inner core anisotropy 内核各向异性 52

ionosphere 电离层 9

iron 铁 57

island arcs 岛弧 73

isotopes 同位素 4

J

JOIDES Resolution "联合果敢"号 60

K

Kelvin, Lord 开尔文勋爵 24

kimberlite 角砾云橄岩 43

Krakatau 喀拉喀托 109

L

Lake Nyos 尼奥斯湖 114
landslides, submarine 海底滑坡 58
laser ranging 激光测距 82
late heavy bombardment 后期重轰炸期 32
Laurasia 劳亚古陆 85
life 生命 15, 31, 56
life, origin of 生命的起源 32
limestone 石灰岩 62
liquefaction 液化 125
lithophile elements 亲岩元素 48
lithosphere 岩石圈 17, 35, 70, 73
Loma Prieta 洛马普列塔 124, 129
Lovelock, James 詹姆斯·洛夫洛克 15
Lyell, Charles 查尔斯·莱尔 23

M

magma chambers 岩浆房 69
magnetic dynamo 磁场发电机 49
magnetic reversals 地磁倒转 50, 68, 82
magnetic stripes 磁条带 67
magnetism 磁力 48
magnetosphere 磁层 6, 7
manganese nodules 锰结核 70
mantle 地幔 17
mantle circulation 地幔循环 73
mantle convection 地幔对流 41, 78

mantle flow 地幔流动 37
mantle plumes 地幔柱 45, 70, 83, 87, 92, 100
mantle pressure 地幔压力 39
mantle, base of 地幔基 44
marine transgression 海侵 59
Mars 火星 95
Martin, John 约翰·马丁 57
Matthews, Drummond 德拉蒙德·马修斯 67
Maunder minimum 蒙德极小期 11
McKenzie, Dan 丹·麦肯齐 87
Mediterranean 地中海 63
melting 熔融 45, 68, 70, 100
Mercalli scale 麦加利震级表 116
metamorphic rocks 变质岩 4
meteorites 陨石 47
methane hydrate 甲烷水合物 63
mid ocean ridge 洋中脊 65, 101
Milankovich cycle 米兰柯维奇循环 11, 63
mineral resources 矿物资源 95
Mohorovicic discontinuity or Moho 莫氏不连续面或莫霍面 35, 38
molasse 磨砾层 91
monsoon 季风 12, 90
Moon 月球 10, 31, 32
Moon, formation of 月球的形成 34
Mount Fuji 富士山 104
Mount Pelée 培雷火山 108

Mount Pinatubo 皮纳图博火山 105
Mount St Helens 圣海伦斯火山 106

N

Nanga Parbat 南伽峰 90
nappe folds 推覆褶皱 92
Neptunist 水成论者 22, 99
Nevado del Ruiz 内瓦多·德·鲁伊斯火山 114
New Madrid 新马德里 121
North Atlantic Ridge 北大西洋洋中脊 65
North Sea 北海 75
nutrients 营养物 56

O

obduction 仰冲作用 74
obsidian 黑曜岩 104
Ocean crust 洋壳 71
ocean currents 洋流 13
ocean drilling programme 大洋钻探计划 15, 60
ocean floor 洋底 58
ocean productivity 海洋生产力 57, 62
ocean trenches 海沟 71
oil 石油 95
olivine 橄榄石 37, 40, 121
orbit 轨道 10

ozone 臭氧 6, 9

P

Pacific 太平洋 71
Pacific ring of fire 太平洋火山带 99
palaeomagnetism 古地磁 87
Pangaea 联合古陆 85
Pangaea Ultima 终极联合古陆 97
peridotite 橄榄岩 37, 40, 100
perovskite 钙钛矿 43, 45
Phillips, John 约翰·菲利普斯 23
photic zone 透光带 55
photosynthesis 光合作用 32
phytoplankton 浮游植物 17
pillow lava 枕状熔岩 65, 69, 74
planets, Earth-like 类地行星 5
plate boundaries 板块边界 73
plate tectonics 板块构造 3, 78, 115
Plinean eruptions 普林尼式火山喷发 103
Pliny the Younger 小普林尼 99
Plutonist 火成论者 99
polar wandering curve 磁极迁移曲线 82
Pompeii 庞贝 109
Pre-Cambrian 前寒武纪 31, 85
pumice 浮岩 104
Pyrenees 比利牛斯山 80
pyroclastic flows 火山碎屑流 108

R

radiation belts 辐射带 7
radio dating 放射性测定地质年龄 24
radioactive decay 放射性衰变 18, 24, 34
rhyolite 流纹岩 105
Richter scale 里克特震级表 116
rock cycle 岩石循环 18
rotation 自转 10
Rutherford, Ernest 欧内斯特·卢瑟福 24

S

San Andreas fault 圣安德烈亚斯断层 118
San Francisco 旧金山 118
Santorini 桑托林岛 109
Scotese, Christopher 克里斯托弗·斯科泰塞 97
sea floor spreading 海底扩张 65, 70
sea level 海平面 59
seamounts 海山脉 58
seismic profiling 地震剖面测量 71
seismic tomography 地震层析成像 38, 73
seismology 地震学 37
Seychelles 塞舌尔群岛 87
shield volcanoes 盾状火山 101
siderophile elements 亲铁元素 47

silicaceous ooze 硅质软泥 62
Smith, William 威廉·史密斯 28
snowball Earth 雪球地球 31
solar system 太阳系 6
solar wind 太阳风 7
sonar 声呐 55
stratovolcanoes 成层火山 104
subduction 潜没 47, 71, 73, 94, 104
sulphide 硫化物 69
sunspots 太阳黑子 11
super-continent 超大陆 74, 85
super-plumes 超级地幔柱 49
suture 地缝合线 74, 94
synthetic aperture radar 合成孔径雷达 119

T

Tambora 坦博拉火山 109
Taylor, Frank 弗兰克·泰勒 78
tectonic plates 构造板块 36, 73
temperature 温度 51
Tangshan, China 中国唐山 127
Tethys ocean 特提斯洋 74, 87, 91
thermosphere 热大气层 9
Tibetan plateau 西藏高原 89
time 时间 21
Tonga trench 汤加海沟 41
transform faults 转换断层 65, 81
tree rings 树木的年轮 26

troposphere 对流层 10

tsunamis 海啸 58

turbidite 浊积岩 59

U

unconformity 不整合 28

uniformitarianism 均变论 23

Ussher, Archbishop 厄谢尔大主教 23

V

van Allen, James 詹姆斯·范艾伦 7

Venus 金星 94

Vesuvius 维苏威火山 99, 104

Vine, Fred 弗雷德·瓦因 67

volcanoes 火山 98—114

W

water 水体 19, 34, 54, 73, 94

water, origin of 水体的形成 55

weather 气候 10

Wegener, Alfred 阿尔弗雷德·魏格纳 3, 78

Western Ghats 西高止山脉 87

Wilson, Tuzo 图佐·威尔逊 78, 85

X

xenoliths 捕虏岩 37, 40

Y

Yucatan Peninsular 尤卡坦半岛 29

Z

zircon 锆石 85, 96

Martin Redfern

THE EARTH
A Very Short Introduction

Contents

Acknowledgements

List of illustrations

1 Dynamic planet 1
2 Deep time 20
3 Deep Earth 35
4 Under the sea 54
5 Drifting continents 76
6 Volcanoes 98
7 When the ground shakes 115

Epilogue 131

Further reading 135

Acknowledgements

The author would like to thank: Arlene Judith Klotzko for introductions without which this book would never have been written; Shelley Cox for initial enthusiasm; Emma Simmons for continued patience; David Mann for instant cartoons; Pauline Newman and Paul Davies for helpful comments; Marian and Edmund Redfern for nurturing my enthusiasm and reading the results; Robin Redfern for beavering away; the un-named readers who have kept me precise; and the countless geologists who have shared with me their time and enthusiasm.

List of illustrations

1 Planet Earth, as seen from *Apollo 17* xiv
Corbis

2 The Earth's magnetic envelope 7

3 The carbon cycle 16

4 'Onion' layers in a radial section of Earth 17

5 The main divisions of geological time 27

6 Circulation in the Earth's mantle 42

7 Possible model for the generation of Earth's magnetic field 50

8 The ocean drilling ship *JOIDES Resolution* 61

9 The global system of ocean ridges 66

10 Stripes of magnetization in ocean floor volcanic rocks 67

11 The principal components of a mid-ocean ridge 68

12 How ocean lithosphere subducts beneath a continent 72

13 The major tectonic plates 79

14 How the continents have changed over the last 200 million years 84

15 Tectonic map of Southeast Asia 88

16 Seismic reflection profile of layers within the Earth's crust 93

17	Types of volcano	102	21	Ground movement in the Izmit earthquake in Turkey 120
18	Eruption of Mount St Helens Popperfoto/United Press International (UK) Limited	107	22	Kobe City, Japan, after an earthquake 123 Sipa Press/Rex Features
19	Features of an erupting composite volcano	111	23	San Andreas fault system, California 124
20	Distribution of major earthquakes in the past 30 years	117		Cartoons © Copyright David Mann

The publisher and the author apologize for any errors or omissions in the above list. If contacted they will be pleased to rectify these at the earliest opportunity.

1. Planet Earth, as seen from *Apollo 17*, December 1972.

Chapter 1
Dynamic planet

> Once a photograph of the Earth, taken from the outside, is available, a new idea as powerful as any in history will be let loose.
>
> Sir Fred Hoyle, 1948

How can you put a big round planet in a small flat book? It is not an easy fit, but there could be two broadly different ways of attempting it. One is the bottom-up approach of geology: essentially, looking at the rocks. For centuries, geologists have scurried around on the surface of our planet with their little hammers examining the different rock types and the mineral grains which make them up. With eye and microscope, electron probe and mass spectrometer, they have reduced the planet's crust to its component parts. Then they have mapped out how the different rock types relate to one another and, through theory, observation, and experiment they have worked out how they might have got there. It has been a huge undertaking and one that has brought deep insights. Collectively, the efforts of all those geologists have built a giant edifice on which future earth scientists can stand. It's as a result of this bottom-up approach that I can write this book. But it is not the approach I will use. This is not a guide to rocks and minerals and geological map-making. It is a portrait of a planet.

The new view on our old planet is the top-down approach of what has come to be known as Earth systems science. It looks at the

Earth as a whole and not just frozen in time in the moment we call now. Taken over the deep time of geology we begin to see our planet as a dynamic system, a series of processes and cycles. We can begin to understand what makes it tick.

The view from above

The prediction above was made by astronomer Sir Fred Hoyle in 1948, a decade before the dawn of space flight. When unmanned rockets took the first pictures of the Earth from outside, and when the first generation of astronauts saw for themselves our world in its entirety, the prediction came true. It's not that those first views told us much we didn't already know about the Earth, but they gave us an icon. And to many of the astronauts who witnessed the view first-hand, it gave an emotional experience of the beauty and seeming fragility of our world that has lived with them ever since. It is perhaps no coincidence that Earth sciences were undergoing their own revolution at the same time. The concept of plate tectonics was at last gaining acceptance, 50 years after Alfred Wegener originally suggested it. Exploration of the ocean floor revealed that it was spreading out from a system of mid-ocean ridges. It had to be going somewhere, forcing continents apart or into one another. The unimaginable masses of continent-sized plates of rock were on the move in an elaborate and ancient waltz.

It was around the same time, and with the same icon of that small blue jewel we call Earth floating in the blackness of space, that a global environmental movement began to form, a mixture of those with a sentimental attachment to endangered species and rainforests and scientists taking on board a new view of complex, interacting ecological systems. Today, most university departments and research groups use the term 'Earth sciences' rather than geology, recognizing a broadening of the discipline beyond the study of rocks. The term 'Earth systems' is becoming widespread, recognizing the inter-related, dynamic nature of processes that include not only the solid, rocky Earth but its oceans, the fragile veil of its atmosphere, and the thin film of life on its surface as well. It is as if our world were an onion; a series of concentric spheres, from magnetosphere and atmosphere, through biosphere and hydrosphere, to the layers of the solid earth. Not all are spherical and some are much less substantial than others, but each manages to persist in a delicate equilibrium. Each component of such a system is seen not as something fixed and unchanging but more like a fountain; maintaining its overall structure perhaps, but constantly changing as material and energy pass through it.

If rocks could talk

Rocks and stones are not the most forthcoming of storytellers. They have a tendency to sit there gathering moss, only rolling when pushed. But geologists have ways of making them talk. They can hit them and slice them; squeeze them, squash them, strain and stress them until they crack – sometimes quite literally. If you know how to look at them, rocks can tell you their history. There is the recent history of the rock on the surface: how it has been weathered and eroded; the tell-tale scars of wind, water, and ice. There can be deeper scars that record periods of heat and pressure and deformation when the rock was buried. Where these changes are extreme the rocks are known as metamorphic. Then there are clues to the origin of the rocks. Some show signs of having once been molten and pushed up from deep within the Earth to erupt out of volcanoes or to intrude into pre-existing rocks. These are the igneous rocks. The size of mineral grains within them can reveal how quickly they were cooled. A large mass of granite cools slowly so that crystals in it are large. Volcanic basalt solidifies rapidly and so is fine-grained. Rocks can be made of the ground-down remains of previous rocks. Here, the size of the fragments tends to reflect the energy of the environment that laid them down: from fine shale and mudstone deposited in still water, through sandstones to coarse conglomerates washed down by raging torrents. Others, such as chalk and limestone, are chemical deposits accumulated as living systems took carbon dioxide from the atmosphere and precipitated it in sea water, turning, as it were, the sky into stone.

Even individual mineral grains have their story to tell. Mineralogists can strip them apart atom by atom in mass spectrometers so sensitive that they can reveal different ratios of isotopes (different atomic forms of the same elements) even among trace constituents. Sometimes these can help date the grains so that we know if they came out of still more ancient rock. They can also reveal the stages of growth of a crystal, for example of diamond, as it passes through the Earth's mantle. In the case of isotopes of carbon and oxygen in minerals derived from marine organisms, it's even possible to estimate the temperature of the sea and the global climate when they formed.

Other worlds

The trouble with the Earth is that it is the only one we've got. We can only see it as it is today, and we can't tell if it is here simply due to some happy accident. That's why Earth scientists are taking a renewed interest in astronomy. Powerful new telescopes sensitive to infrared and sub-millimetre wavelength radiation can stare deep into star-forming regions to see what may have happened when our own solar system was born. Around some of the young stars they have revealed dusty haloes known as proto-planetary discs, perhaps new solar systems in formation. But the search for fully formed Earth-like planets is more difficult. To see directly such a planet in orbit around a distant star would be like trying to spot a small moth close to a powerful searchlight. But indirect methods have led to the discovery of planets in recent years, mainly by detecting tiny wobbles in the motions of the parent stars due to gravitational effects. The clearest effects and therefore the first discovered seem to be due to planets far bigger than Jupiter orbiting far closer to their stars than the Earth is to the Sun. So they could hardly be termed Earth-like. But evidence is beginning to accumulate for solar systems more like our own, with multiple planets. Though small and hospitable planets like Earth will be hard to detect.

To see such planets directly would take telescopes in space that we can scarcely dream of. There are ambitious plans underway in both the USA and Europe for a network of linked infrared telescopes. Each would have to be far bigger than the Hubble Space Telescope, and four or five of them would have to fly in close formation to combine their signals to resolve the planet. They would have to be as far out as Jupiter to get beyond the dusty infrared glow of our own planetary system. But then, they might be able to detect vital signs in distant planetary atmospheres and, in particular, they might detect ozone. That would imply Earth-like conditions of climate and chemistry plus the existence of free oxygen, something which, as far as we know, can only be maintained by life.

Signs of life

In February 1990, on its way out of the solar system after encounters with Jupiter and Saturn, the Voyager I probe beamed back the first image of our entire solar system as it might appear to visitors from another star. The picture is dominated by a single bright star, our Sun, seen from 6 billion kilometres away, 40 times the distance from which we are used to seeing it. The planets are scarcely visible. The Earth itself is smaller than one picture element in Voyager's camera, its faint light caught in what looks like a sunbeam. This is our whole world, seemingly just a speck of dust. But to any alien visitor with the right instruments, that tiny blue world would immediately attract attention. Unlike the giant stormy gas bags of the outer planets, cold, dry Mars, or the acid steam-bath of Venus, the Earth has everything just right. Water exists in all three phases – liquid, ice, and steam. The atmospheric composition is not that of a dead world that has reached equilibrium but one that is active and must be constantly renewed. There is oxygen, ozone, and traces of hydrocarbons; things that would not exist together for long if they were not constantly renewed by living processes. This alone would attract the attention of our alien visitors, even if they could not detect the constant babble of our communications, radio and television.

Magnetic bubble

Geophysics goes way above our heads. I don't mean by that that it is incomprehensible but that the physical influence of our planet extends far above its solid surface, way out into what we regard as empty space. But it is not empty. We live in a series of bubbles nested like Russian dolls one within another. The Earth's sphere of influence lies within the greater bubble dominated by our Sun. That in turn lies within overlapping bubbles blown by the expanding debris of exploding stars or supernovae, long, long ago. They are all within our Milky Way galaxy, which is in turn a member of a supercluster of galaxies within the known universe, which itself may be a bubble in a quantum foam of worlds.

The Earth's atmosphere and magnetic field shield us, for the most part, from the radiation hazards from space. Without this protection, life on the Earth's surface would be threatened by solar ultraviolet and X-rays as well as cosmic rays, high-energy particles from violent events throughout the galaxy. There is also a permanent gale of particles, mostly hydrogen nuclei or protons, blowing outwards from the Sun. This solar wind speeds past the Earth at typical velocities of around 400 kilometres per second, and goes three times faster during a solar storm. It extends for billions of kilometres out into space, beyond all the planets and maybe beyond the orbits of comets, which reach out many thousands of times further from the Sun than does the Earth. The solar wind is very tenuous but it is sufficient to blow out the tails of comets as they come closer in to the heart of the solar system, so the tails always point away from the Sun. It also features in imaginative proposals for propelling spacecraft with vast gossamer-thin solar sails.

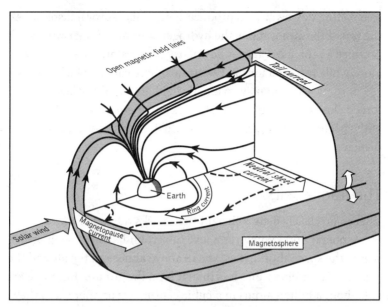

2. Diagram of the Earth's magnetic envelope, the magnetosphere, swept back into a comet-like structure by the solar wind. Arrows show the directions of electrical currents.

The Earth is sheltered from the solar wind by its magnetic field, the magnetosphere. Because the solar wind is electrically charged it represents an electrical current, which cannot cross magnetic field lines. Instead, it compresses the Earth's magnetosphere on the sunward side, like the bow wave of a ship at sea, and stretches it out into a long tail down-wind which reaches almost as far as the orbit of the Moon. Charged particles caught within the magnetosphere build up in belts between the field lines where they are forced to spiral, generating radiation. These radiation belts were first spotted in 1958 when James van Allen flew the first Geiger counter in space on board the American Explorer 1 satellite. They are areas to be avoided by spacecraft hoping for a long life and would be lethal to unprotected astronauts.

Where the Earth's magnetic field lines dive down towards the poles, solar wind particles can enter the atmosphere, sending atoms ricocheting downwards to produce spectacular auroral displays. At the top of the atmosphere, the hydrogen ions of the solar wind itself produce a pink haze. Lower down, oxygen ions produce a ruby-red glow, while nitrogen ions in the stratosphere cause violet blue and red auroras. Occasionally, magnetic field lines in the solar wind are forced close to those of the Earth, causing them to reconnect, often with spectacular releases of energy which extend the auroral displays.

The fragile veil

There's no clearly defined height that marks the top of the atmosphere; 260 kilometres above the ground, in low Earth orbit where the space shuttle flies, you're above almost all the air and the pressure is a billion times less than it is on the ground. But there are still about a billion atoms in a cubic centimetre up there, and they are hot and electrically charged and hence can have a corrosive effect on space vehicles. At times of maximum solar activity, the atmosphere expands slightly, exerting more frictional drag on low spacecraft, which have to be boosted up to stay in orbit. The upper atmosphere, above 80 kilometres, is sometimes known as the thermosphere because it is so hot, even though it is so rarefied that you would not burn your skin on it.

This region of the atmosphere also absorbs dangerous X-rays and some of the ultraviolet radiation from the Sun. As a result, many atoms become 'ionized', that is they lose an electron. For this reason, the thermosphere is also called the ionosphere. Because the ionosphere is electrically conducting, it will reflect certain frequencies of radio waves, making it possible for short-wave radio transmissions to be heard around the world, well over the horizon from the transmitter.

Even a mere 20 kilometres up, below the thermosphere, the mesosphere, and most of the stratosphere, we are still above 90% of the air in the atmosphere. It is at around this height that we encounter the tenuous ozone layer, molecules containing three oxygen atoms. Ozone forms when ordinary oxygen molecules of two atoms are split by solar radiation and some recombine in threes. Ozone is a highly effective sunscreen for the planet. If all the ozone in the Earth's atmosphere were concentrated at ground level, it would form a layer only about three millimetres thick. But it still filters out virtually all of the most dangerous short-wave UV C radiation from the Sun and most of the medium-wavelength UV B rays as well. Thus it protects life from sunburn and skin cancer. The ozone layer has been severely depleted by chemicals such as CFCs (chlorofluorocarbons) released by human activity, leading to a generalized thinning of the layer and more specific holes over polar regions in the still, cold air of spring. International agreements have slowed the release of CFCs and the ozone layer should recover, but the chemicals are long-lived and it will be some time yet before it does.

Circles and cycles

It is in the lowest 15 kilometres of the atmosphere, the troposphere, that most of the action takes place. This is where weather happens. It's where clouds form and disperse and where winds blow, transferring heat and moisture around the planet. In a dynamic planet, everything seems to go round in circles, flows of energy. And here, close to the surface, these cycles are driven by solar power. There are the obvious cycles of day and night as the Earth spins on its axis and the ground alternately heats and cools, and the annual cycle of the seasons as the Earth orbits the Sun, presenting first more of one hemisphere then more of the other to the sunshine. But there are longer cycles too, such as the wobble of the Earth's axis over tens of thousands of years.

Just as the Earth orbits the Sun, so the Moon orbits the Earth. It takes about 28 days to complete an orbit, giving us our months. As the Earth spins on its axis, the Moon's gravity pulls a bulge in the oceans around the planet, creating tides. This also acts as a brake on the rotation of the Earth, slowing down the days. Daily growth bands in fossil corals 400 million years old suggest that their days were several hours shorter than our own.

The Moon helps to stabilize the orbit of the Earth and hence the climate. But there are far longer cycles at work too. The Earth's orbit around the Sun is not a perfect circle, but an ellipse, with the Sun at one focus. Hence the distance of the Earth from the Sun varies during its orbit. In addition, the degree of variation itself changes over a 95,800-year period. Also the Earth's rotation axis slowly wobbles or precesses like a spinning top off balance. Over a period of 21,700 years the planet's axis traces out a complete cone. At present, the Earth is nearest to the Sun during the northern hemisphere winter. The inclination of the Earth's spin axis with that of its orbit around the Sun (the obliquity) also changes on a 41,000-year period. These so-called Milankovich cycles add up over tens or hundreds of thousands of years to affect climate. They have been blamed for such phenomena as the ice ages that have affected the Earth over the last three and a quarter million years. But the reality is probably even more complex, with their effects amplified or reduced by factors such as ocean circulation, cloud cover, atmospheric composition, volcanic aerosols, the weathering of rocks, biological productivity, and so on.

Solar cycles

Cycles of change are not restricted to the Earth. The Sun can change too. Over its 5-billion-year history, the Sun has been getting progressively warmer. Surface temperatures on Earth have remained more constant, however, as levels of greenhouse gases have fallen over the same time. This was largely due to the effects of life, as plants and algae consumed carbon dioxide that acted like a blanket to keep the young Earth warm. There have been other solar variations as well. There is a regular solar cycle of 11 years which sees a rise and fall in sunspot activity, in turn reflecting the cycle of solar magnetic activity which produces storms and the solar wind. Other Sun-like stars seem to spend about a third of their time free of sunspots, a state called a Maunder minimum. That happened to our Sun between 1645 and 1715 AD. Solar power only dropped by about 0.5%, but this was enough to plunge northern Europe into what has become known as the Little Ice Age, with a series of particularly severe winters. The River Thames in London froze over, and markets and frost fairs were held on it.

Hot air

The Sun distributes its warmth unevenly, warming up equatorial regions the most. As the air warms it tries to expand, increasing atmospheric pressure. To try to restore equilibrium, winds begin to blow and the air circulates. Whilst all this goes on, the Earth continues to rotate, giving the air angular momentum. That is greatest at the equator and results in the so-called Coriolis effect. The atmosphere is not firmly coupled to the solid planet, so, as winds blow away from the equator, they have a momentum that is independent of the rotating surface beneath. This means that, relative to the surface, the winds curve to the right in the northern hemisphere and left in the south. This leads to rotating systems of high and low air pressure, the weather systems that bring us rain or sunshine.

Land masses and mountain ranges influence the circulation of heat and moisture too. Until the Himalayan mountain range began to rise, there was no Indian monsoon, for example. And most importantly, the oceans play a huge part in storing heat and transporting it around the globe. The top 2 metres of the ocean have the same heat capacity as the entire atmosphere. At the same time, heat circulates in ocean currents. But currents on the surface are only half the picture. A good example is the Gulf Stream in the North Atlantic. That carries warm water north and east from the Gulf of Mexico and is one of the reasons why the climate of northwest Europe is much milder in winter than that of northeast America. As the warm water heads north, some evaporates into the clouds, which always seem to fall on British holidaymakers. The remaining surface waters in the ocean cool and become progressively more salty. As a result, they also become denser and eventually sink down to flow back south in the deep Atlantic, completing the conveyor belt of the ocean circulation.

Sudden freeze

About 11,000 years ago, the Earth was emerging from the last Ice Age. Ice was melting, sea levels were rising, and the climate was getting generally warmer. Then, suddenly, in the space of a few years, it turned cold again. The change was particularly marked in Ireland, where pollen in sediment cores shows that the vegetation suddenly reverted from temperate woodland back to tundra dominated by a little plant called the Dryas. Wally Broecker of the Lamont Doherty Geological Observatory has worked out what may have happened. As the ice sheet over North America receded, a vast lake of fresh meltwater, far bigger than the present Great Lakes, built up in central Canada. At first, it drained over a great ridge of rock into the Mississippi. As the ice receded, it suddenly opened a far lower passage down the St Lawrence River to the east. The vast lake of cold fresh water drained into the North Atlantic almost instantaneously. So much water was involved that it caused an immediate sea level rise of 30 metres. It diluted the salty surface water of the North Atlantic and put a virtual stop to the conveyor belt of ocean circulation. Thus there was no warming current into the North Atlantic and Arctic conditions returned. A thousand years later, the ocean circulation resumed as quickly as it had vanished and a temperate climate returned.

The North Atlantic deep water, together with cold bottom water from the Antarctic, finds its way at depth as far as the Indian and

Pacific oceans. The deep current continues into the North Pacific, slowly accumulating nutrients as it goes, before it rises again to the surface.

Global greenhouse

Some of the gases in the Earth's atmosphere act rather like the glass of a greenhouse, letting sunlight in to warm the ground but then preventing the resulting infrared heat radiation from escaping. Were it not for the greenhouse effect, average global temperatures would be around 15 degrees Celsius lower than they are, making life almost impossible. The principal greenhouse gas is carbon dioxide but others, including methane, play an important role. So does water vapour, an effect that is sometimes forgotten. Over hundreds of millions of years, an approximate balance has been struck with plants removing carbon dioxide from the atmosphere through photosynthesis and animals returning it by respiration. Vast quantities of carbon have been buried in sediments such as limestone, chalk, and coal. Volcanic eruptions have released carbon from inside the Earth.

In recent years, concern has grown over what should be termed the enhanced greenhouse effect, the very significant rise in greenhouse gas levels in the atmosphere resulting from human activity. The burning of fossil fuels such as coal and oil are prime culprits, but so are agricultural practices which produce methane, and deforestation which releases carbon dioxide from timber and soils as well as reducing the plant cover to absorb it again. Climate models suggest that these activities could result in a global temperature rise of several degrees over the next century, accompanied by greater extremes in weather and a possible sea level rise.

Climate change

The steady annual rise in carbon dioxide levels has been carefully recorded from an isolated mountain peak in Hawaii since 1958. More than 130 years of meticulous weather readings around the world confirm an average global warming of about half a degree, with the effect particularly pronounced over the last 30 years. But natural climate records go back much, much further. Tree rings record periods of drought and severe frosts, as well as the frequency of wild fires, over their lifespan. Extrapolating overlapping sequences in preserved timber can reveal climate conditions back to 50,000 years ago. Coral growth rings reveal sea surface temperatures over a similar span. Pollen grains in sediments record shifts in vegetation patterns over 7 million years. Landscapes reveal past glaciation and changes in sea level over billions of years. But some of the best records come from cores drilled from ice and from ocean sediments. Ice cores reveal not only rates of snow accumulation and trapped volcanic dust, but bubbles in the ice represent samples of the ancient atmosphere trapped in the snow. Isotopes of hydrogen, carbon, and oxygen can also indicate global temperature at the time. The ice record from Antarctica and Greenland now goes back over 400,000 years. Marine sediments all around the planet have been sampled by the ocean drilling programme and can carry records up to 180 million years old. Isotope ratios in microfossils trapped in the sediments can reveal temperature, salinity, atmospheric carbon dioxide levels, ocean circulation, and the extent of polar ice caps. All these different records reveal that climate change is a fact of life and that long periods in the past have been considerably warmer than the climate we experience today.

Web of life

The most insubstantial of the Earth's layers has had perhaps the most profound effect on the planet: life. Without life, the Earth might be a runaway greenhouse world like Venus, or possibly a cold desert like Mars. There would certainly not be the temperate climate and oxygen-rich atmosphere in which we flourish. We've already heard how the first algae kept pace with the warming Sun by eating the carbon dioxide blanket that insulated the young Earth. The independent scientist James Lovelock suggests that such feedback mechanisms have managed the terrestrial climate for more than 2 billion years. He uses the term Gaia, after the ancient Greek earth goddess, as a name for this system. He does not pretend that there is anything conscious or deliberate about this control; Gaia does not have divine powers. But life, principally in the form of bacteria and algae, does play a key role in the homeostatic process that keeps the planet habitable. A simple computer model called 'Daisyworld' shows how two or more competing species can set up a negative feedback system that controls the environment within habitable limits. Lovelock suspects that the global system on Earth will adapt as human activity enhances the greenhouse effect, even if the adaptations are not favourable to human life.

The carbon cycle

Carbon is forever moving around. Each year, roughly 128 billion tonnes is released as carbon dioxide into the atmosphere by processes on land, and nearly as much is immediately absorbed again by plants and by the weathering of silicate rocks. At sea, the figures are comparable, though slightly more goes in than comes out. The system would be more or less in balance were it not for volcanic emissions and the 5 billion tonnes released each year by burning fossil fuels. The total amount of carbon held in the atmosphere is quite small – just 740 million tonnes, only slightly more than that held in plants and animals on land and slightly less than that held by living things in the ocean. By comparison, the amount of carbon stored in solution in the oceans is vast at 34 billion tonnes, and the amount stored in sediments is 2,000 times greater still. So the physical processes of solution and precipitation may be even more important in the carbon cycle than the biological ones. But life seems to hold some key cards. Carbon incorporated by phytoplankton would be released back to sea water and hence the atmosphere very quickly were it not for the physical properties of copepod faecal pellets. These tiny planktonic animals excrete their waste in small, dense pellets which can slowly sink into the deep ocean, removing them, at least temporarily, from the cycle.

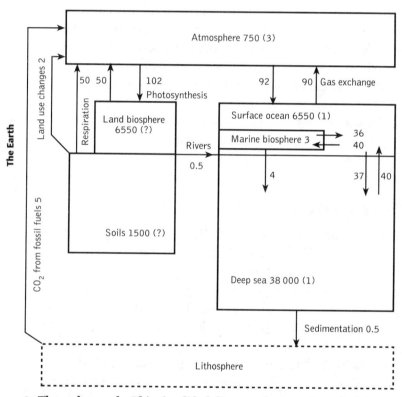

3. The carbon cycle. This simplified diagram shows estimates of the amount of carbon (in billions of tonnes) stored in the atmosphere, oceans, and land. Figures by the arrows show the annual fluxes between stores, figures in brackets the annual net rise. Though small compared with most other fluxes, the input from burning fossil fuels seems enough to upset the balance.

Almost an onion

The interior of the Earth is rather like an onion, made up of a series of concentric shells or layers. On the top is a crust, averaging 7 kilometres thick under the ocean and 35 kilometres thick in continents. That sits on the hard, rocky lithosphere at the top of the mantle, and below it is the softer asthenosphere. The upper mantle extends to a depth of about 670 kilometres, the lower mantle goes down to 2,900 kilometres. Below a thin transition layer comes the liquid outer core of molten iron and a solid iron inner core about the size of the Moon. But it is not a perfect onion. There are horizontal differences within layers, variations in thickness of layers, and, we now know, continuous exchange of material between layers. Where our planet departs from the perfect onion model is where most off the interest and excitement in modern geophysics

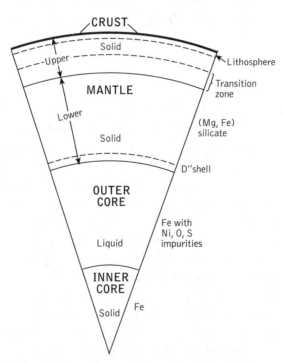

4. The main 'onion' layers in a radial section of the Earth.

lies, and where we can find the clues to the processes that drive the system.

Lava lamps

Do you remember those lava lamps of the 1960s and numerous later revivals? They make a good model for the processes at work within the Earth. Whilst they are switched off, a layer of red gloop sits at the bottom of a layer of transparent oil. But turn on the lamp and the filament in its base warms the red gloop so it expands, becomes less dense, and begins to rise in elongated lumps to the top of the oil. When it has cooled sufficiently, it sinks back down. So it is in the Earth's mantle. Heat from radioactive decay and from the Earth's core drives a sort of heat engine in which the not-quite-solid rock of the mantle slowly circulates over billions of years. It is this circulation that drives plate tectonics, causes the continents to drift, and triggers volcanoes and earthquakes.

The rock cycle

At the surface the results of that heat engine beneath our feet and the solar furnace above our heads meet and drive the rocks full circle. Mountain ranges lifted high by mantle circulation and continental collisions are weathered down by solar-powered wind, rain, and snow. Chemical processes are at work as well. Oxidation by the atmosphere and chemical dissolution by acids from living organisms and dissolved gases help to break down the rocks. Large quantities of carbon dioxide can dissolve in rainwater to make a weak acid that causes chemical weathering, turning silicate minerals into clay. These remains are washed back down to estuaries and oceans where they form new sediments, eventually to be scooped up into new mountain ranges or carried back down into the mantle for deep recycling. The whole process is lubricated by water incorporated into the crystal structure of minerals. This rock cycle was first suggested in the 18th century by James Hutton, but then he had no idea of the depths and the time scales over which it occurs.

So far we have glimpsed just the surface of our amazing planet. Now we will dig deeper into the rocks and into time.

Planetary data	
Equatorial diameter	12,756 km
Volume	1.084×10^{12} km^3
Mass	5.9742×10^{24} kg
Density	5.52 of water
Surface gravity	9.78 m s^{-2}
Escape velocity	11.18 km s^{-1}
Day length	23.9345 hrs
Year length	365.256 days
Axial inclination	23.44°
Age	4,600 million years approx.
Distance from the Sun	Min. 147 million km
	Max. 152 million km
Surface area	5,096 million km^2
Land surface	148 million km^2
Ocean cover	71% of surface
Atmosphere	N$_2$ 78% O$_2$ 21%
Continental crust	35 km thick, average
Oceanic crust	7 km thick, average
Lithosphere	To 75 km depth
Mantle (silicates)	2,900 km thick
	About 3,000 °C at base
Outer core (molten iron)	2,200 km thick
	About 4,000 °C at base
Inner core (solid iron)	1,200 km thick
	Up to 5,000 °C at centre

Chapter 2
Deep time

Space is big, really big ... You may think it's a long way down the street to the chemists' but that's just peanuts to space.
        Douglas Adams, *The Hitchhiker's Guide to the Galaxy*

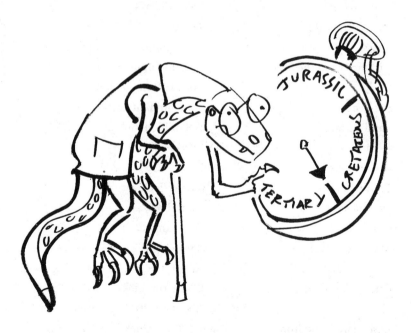

The world is not only large in its spatial dimensions. It also extends almost unimaginably far back in time. It is impossible to get a full grasp of the concepts and processes at work in geology without an understanding of what writers John McPhee, Stephen Jay Gould, and Henry Gee have referred to as deep time.

Most of us know our parents, many remember our grandparents. Only a few have met great grandparents. Their youth lies more than a century in our past, a time which seems alien to us with our vastly different scientific understanding and social structure. Just a dozen generations back, England was ruled by Queen Elizabeth I, motorized transport and electronic communication was undreamt of, and Europeans were exploring the Americas for the first time. Thirty generations takes us a thousand years back, before the Normans invaded Britain. It is also before continuous written records are likely to be able to trace our direct ancestry. We may be able to tell from archaeology and genetics roughly who our ancestors were at this time and where they might have lived but we cannot be certain. Fifty generations ago, the Roman Empire was in full swing. And 150 generations back, the Great Pyramid of Ancient Egypt had not been constructed. About 300 generations takes us back to the Neolithic in Europe at a time when the last Ice Age had only just ended and simple agriculture was the latest technological revolution. It is unlikely that archaeology can reveal where our ancestors were living at that time, though comparisons of our maternally inherited mitochondrial DNA may indicate the broad region. Add another zero to the year and we have gone back 3,000 generations to 100,000 years ago. At this time, we cannot trace separate ancestry of any living racial group. Mitochondrial DNA suggests that there was a single maternal ancestor of all modern humans in Africa not long before. But, in geological time, this is still recent.

Ten times older at a million years and we start to lose track of the modern human species. Another factor of ten and we are looking at the fossil remains of early ape ancestors. This far back it's impossible to point even to a single species and say with certainty that amongst these individuals was our ancestor. Multiply by ten again and, 100 million years ago, we are in the age of the dinosaurs. The ancestor of humans must be some insignificant shrew-like creature. A thousand million years ago and we are back amongst the first fossils, maybe before even the first recognizable animals. Ten billion years ago and we are before the birth of the Sun and solar system, at a time when the atoms that today make up our planet and ourselves were being cooked in the nuclear furnaces of other stars. Time is indeed deep.

Now think again of the changes that can take place in a few generations. Historical time is trivial compared to the age of the Earth, yet a few centuries have seen many volcanic eruptions, cataclysmic earthquakes, and devastating landslides. And think of the relentless progress of less devastating changes. In 30 generations, parts of the Himalayas have risen by maybe a metre or more. But at the same time they have eroded, probably by more than this. Islands have been born, others washed away. Some coasts have eroded back hundreds of metres, others have been left high and dry. The Atlantic has widened by about 30 metres. Now multiply all these comparatively recent changes by factors of ten or a hundred or a thousand, and you are beginning to see what can happen over geological 'deep time'.

Flood and uniformity

Humans have noticed fossil remains since prehistoric times. There are ancient stone tools that appear to have been chipped so as to show off a fossil shell. The fossilized stem of a giant cycad was placed in an ancient Etruscan burial chamber. But attempts to understand the nature of fossils are comparatively recent. The science of geology arose primarily in Christian Europe where beliefs based on biblical stories made it unsurprising to discover the shells and bones of extinct creatures high up in mountainous regions: they were the remains of animals that perished in the biblical flood. Even granite, it was suggested by so-called neptunists, was precipitated from an ancient ocean. The idea of extreme acts of God such as the flood helped people to imagine that the Earth had been shaped by catastrophes, and this was the generally accepted theory until the end of the 18th century.

In 1795 the Scottish geologist James Hutton published his now famous *Theory of the Earth*. The much quoted though paraphrased summary of its message is that 'the present is the key to the past'. This is the theory of gradualism or uniformitarianism, which says that if you want to understand geological processes you must look at the almost imperceptibly slow changes occurring today and then simply trace them out through history. It was a theory developed and championed by Charles Lyell, who was born in 1797, the year Hutton died. Both Hutton and Lyell tried to put religious beliefs in events such as the creation and the flood to one side and proposed that the gradual processes at work on the Earth were without beginning or end.

Dating creation

Attempts to calculate the age of the Earth came originally out of theology. It is only comparatively recently that so-called creationists have interpreted the Bible literally and therefore believe that Creation took just seven 24-hour days. St Augustine had argued in his commentary on Genesis that God's vision is outside time and therefore that each of the days of Creation referred to in the Bible could have lasted a lot longer than 24 hours. Even the much quoted estimate in the 17th century by Irish Archbishop Ussher that the Earth was created in 4004 BC was only intended as a minimum age and was based on carefully researched historical records, notably of the generations of patriarchs and prophets referred to in the Bible.

The first serious attempt to estimate the age of the Earth on geological grounds was made in 1860 by John Phillips. He estimated current rates of sedimentation and the cumulative thickness of all known strata and came up with an age of nearly

96 million years. William Thompson, later Lord Kelvin, followed this with an estimate based on the time it would have taken the Earth to cool from an originally hot molten sphere. Remarkably, the first age he came up with was also very similar at 98 million years, though he later refined it downwards to 40. But such dates were considered too recent by uniformitarianists and by Charles Darwin, whose theory of evolution by natural selection required more time for the origin of species.

By the dawn of the 20th century, it had been realized that additional heat might come from radioactivity inside the Earth and so geological history, based on Kelvin's idea, could be extended. In the end, however, it was an understanding of radioactivity that led to the increasingly accurate estimates of the age of the Earth that we have today. Many elements exist in different forms, or isotopes, some of which are radioactive. Each radioactive isotope has a characteristic half-life, a time over which half of any given sample of the isotope will have decayed. By itself, that's not much use unless you know the precise number of atoms you start with. But, by measuring the ratios of different isotopes and their products it is possible to get surprisingly accurate dates. Early in the 20th century, Ernest Rutherford caused a sensation by announcing that a particular sample of a radioactive mineral called pitchblende was 700 million years old, far older than many people thought the Earth to be at that time. Later, Cambridge physicist R. J. Strutt showed, from the accumulation of helium gas from the decay of thorium, that a mineral sample from Ceylon (now Sri Lanka) was more than 2,400 million years old.

Uranium is a useful element for radio dating. It occurs naturally as two isotopes – forms of the same element that differ only in their number of neutrons and hence atomic weight. Uranium-238 decays via various intermediaries into lead-206 with a half-life of 4,510 million years, whilst uranium-235 decays to lead-207 with a 713-million-year lifetime. Analysis of the ratios of all four in rocks, together with the accumulation of helium that comes from the

Some radioisotopes used for dating

Isotope	Product	Half-life	Use
Carbon 14	Carbon 12	5,730 years	Dating organic remains up to 50,000 years
Uranium 235	Lead 207	704 million years	Dating intrusions and individual mineral grains
Uranium 238	Lead 206	4,469 million years	Dating individual mineral grains in ancient crust
Thorium 232	Lead 208	14,010 million years	As above
Potassium 40	Argon 40	11,930 million years	Dating volcanic rocks
Rubidium 87	Strontium 87	48,800 million years	Dating granitic igneous and metamorphic rocks
Samarium 147	Neodymium 143	106,000 million years	Dating basaltic rocks and very ancient meteorites

decay process, can give quite accurate ages and was used in 1913 by Arthur Holmes to produce the first good estimate of the ages of the geological periods of the past 600 million years.

The success of radio-dating techniques is due in no small way to the power of the mass spectrometer, an instrument which can virtually sort individual atoms by weight and so give isotope ratios on trace constituents in very small samples. But it is only as good as the assumptions that are made about the half-life, the original abundances of isotopes, and the possible subsequent escape of decay products. The half-life of uranium isotopes makes them good for dating the earliest rocks on Earth. Carbon 14 has a half-life of a mere 5,730 years. In the atmosphere it is constantly replenished by the action of cosmic rays. Once the carbon is taken up by plants and the plants die, the isotope is no longer replenished and the clock starts ticking as the carbon 14 decays. So it is very good for dating wood from archaeological sites, for example. However, it turns out that the amount of carbon 14 in the atmosphere has varied along with cosmic ray activity. It is only because it has been possible to build up an independent chronology by counting the annual growth rings in trees that this came to light and corrections to carbon dating of up to 2,000 years could be made.

The geological column

Look at a section of sedimentary rocks in, for example, a cliff face and you will see that it is made up of layers. Sometimes annual layers corresponding to floods and droughts are visible. More often, the layers represent occasional catastrophic events or slow but steady sedimentation across hundreds of thousands or even millions of years, followed by a change of environment leading to a layer of slightly different rock. In the case of a really deep section of ancient rock, such as that seen in the Grand Canyon in Arizona, hundreds of millions of years of deposits are represented. It is a natural human instinct to divide up and classify things, and sedimentary rock with its many layers is an obvious candidate. But,

Main divisions of geological time

Eon	Era	Period		Epoch	Age
Phanerozoic	Cenozoic	Quaternary		Holocene	0.01
				Pleistocene	1.8
		Tertiary	Neogene	Pliocene	5.3
				Miocene	23.8
			Palaeogene	Oligocene	33.7
				Eocene	54.8
				Palaeocene	65.0
	Mesozoic	Cretaceous		UPPER	
				LOWER	142
		Jurassic		UPPER	
				MIDDLE	
				LOWER	205.7
		Triassic		UPPER	
				MIDDLE	
				LOWER	248.2
	Palaeozoic	Permian		UPPER	
				LOWER	290
		Carboniferous	Pennsylvanian	UPPER	323
			Mississippian	LOWER	354
		Devonian		UPPER	
				MIDDLE	
				LOWER	417
		Silurian		UPPER	
				LOWER	443
		Ordovician		UPPER	
				MIDDLE	
				LOWER	495
		Cambrian		UPPER	
				MIDDLE	
				LOWER	545
Precambrian	Proterozoic				2500
	Archaean				4000
	Hadean				4560

5. The main divisions of geological time (not to scale). Ages (on the right, in millions of years before present) are those agreed by the International Commission on Stratigraphy in 2000.

when viewing a spatially narrow cliff face of flat layers, it's easy to forget that the layers are not continuous around the world. The entire globe was never covered by a single shallow ocean depositing similar sediments! Just as today, there are rivers, lakes, and seas, deserts, forests, and grasslands, so in ancient times there was a panoply of sedimentary environments.

It was an English civil engineer, William Smith, who, in the early 19th century, began to make sense of it all. He was surveying for Britain's new canal network and started to realize that rocks in different parts of the country sometimes contained similar fossils. In some cases the rock types too were the same, sometimes only the fossils were similar. This enabled him to correlate the rocks in different places and work out an overall sequence. As a result, he published the first geological map. Once the dates were added in the 20th century, and the rocks correlated between different continents, it was possible to publish a single sequence of layers representing periods of geological time for the whole world. The geological column we know today is the product of many techniques, refined over the years and agreed by international collaboration.

Extinctions, unconformities, and catastrophes

It became clear that some of the changes in the geological column were bigger than others, and these provided convenient places to divide the geological past into separate eras, periods, and epochs. Sometimes there was a sudden and significant change in the nature of the rocks across such a boundary, indicating a major environmental change. Sometimes there was what is known as an unconformity, a break in deposition, caused, for example, by a change in sea level so that either deposition stopped or the layers were eroded away before the column continued. They are often also marked by major changes in fauna, represented by fossils, with many species becoming extinct and new ones beginning to arise.

A few intervals in the geological record stand out for the severity of the extinctions across them. The end of the Cambrian period and the end of the Permian period were both marked by the extinctions of around 50% of families and up to 95% of individual species of marine invertebrates. The extinctions that marked the late Triassic and late Devonian saw the loss of about 30% of families and, slightly smaller at 26%, but the most recent and the most famous, is the mass extinction at the end of the Cretaceous period 65 million years ago. That so-called K/T boundary is famous not only because it saw the extinction of the last of the dinosaurs but also because there is good evidence for the cause.

Threat from space

The first suggestion, by Walter and Louis Alvarez, that the extinction might be due to an astronomical impact at first received little scientific support. However, they soon discovered that sediments in a narrow band at that point in the geological column were enriched in iridium, an element abundant in some types of meteorite. But there was no sign of an impact crater of that age. Then evidence began to emerge, not from the land but from the sea just off the Yucatan Peninsula of Mexico, of a buried crater 200 kilometres across. There is evidence of debris from a much wider area. If, as is calculated, it marks the point where an asteroid or comet, maybe 16 kilometres across, hit the Earth, the results would indeed have been devastating. Apart from the effects of the impact itself and the tsunami that resulted, so much rock would have been vaporized that it would have spread round the Earth in the atmosphere. At first it would have been so hot that its radiant heat would have triggered forest fires on the ground. The dust would have stayed in the atmosphere for several years, blocking out sunlight, creating a global winter, and causing food plants and plankton to die. The sea bed at the impact site included rocks rich in sulphate minerals and these would have vaporized, leading to a deadly acid rain when it washed out of the atmosphere again. It is almost surprising that any living creatures survived.

The menace within

It was once hard to understand how any mass extinctions could have occurred. Now, there are so many competing theories that it is difficult to choose between them. They mostly involve severe climate change, whether triggered by a cosmic impact, changing sea levels, ocean currents and greenhouse gases, or a cause from within the planet such as rifting or major vulcanism. It does seem that most of the mass extinctions we know coincided at least approximately with major eruptions of flood basalts. In the case of the late Cretaceous, it was the eruptions that produced the Deccan Traps in western India. There has even been a suggestion that a major asteroid impact caused shock waves to focus on the other side of the Earth, triggering eruptions. But the times and positions do not seem to line up well enough to prove that explanation. Whatever the reason, the history of life and of the planet has been punctuated by some catastrophic events.

Chaos reigns

We can all remember climatic events that stand out, say over the last decade, as the worst winter, flood, storm, or drought. Take the record back for a century and the likelihood is that an even bigger one will stand out. Authorities often use the concept of a '100-year' flood in planning coastal or river flood defences; they are designed to withstand the sort of flood that only happens once a century. It's likely to be more severe than the sort which happens only once a decade. But, if you extend the same idea to a thousand years or a million years, there is likely to be one that will be bigger still. According to some theorists, that is likely to be true of anything from floods, storms, and droughts to earthquakes, volcanic eruptions, and asteroid impacts. Over geological time we had better watch out!

Deeper time

The list of geological periods that is often shown in books goes back only about 600 million years to the start of the Cambrian period. But that ignores 4 billion years of our planet's history. The trouble with most Pre-Cambrian rocks is that they are, as Professor Bill Schopf of the University of California puts it, fubaritic – fouled up beyond all recognition. The constant tectonic reprocessing of the Earth from within, and the relentless pounding of weather and erosion from above, mean that most of the Pre-Cambrian rocks that survive at all are deeply folded and metamorphosed. But on most clear nights you can see rocks that are more than 4 billion years old – by looking up at the Moon rather than down at the Earth. The Moon is a cold, dead world with no volcanoes and earthquakes, water or weather to resurface it. Its surface is covered with impact craters, but most of those happened early in its history when the solar system was still full of flying debris.

The Pre-Cambrian rocks that do survive on Earth tell a long and fascinating story. They are not, as Darwin had supposed, devoid of the traces of life. Indeed, the end of the Pre-Cambrian, from about 650 to 544 million years ago, has yielded a rich array of strange fossils, particularly from localities in southern Australia, Namibia, and Russia. Prior to that there seems to have been a particularly severe period of glaciation. The phrase 'snowball Earth' has been used, conveying the possibility that all the world's oceans froze over. Inevitably, that would have been a major setback for life, and there is scant evidence for multicellular life forms before this. But there is abundant evidence for microorganisms – bacteria, cyanobacteria, and filamentous algae. There are filamentous microfossils from Australia and South Africa that are around 3,500 million years old, and there is what looks like the chemical signature of life in carbon isotopes in rocks from Greenland that are 3,800 million years old.

During the first 700 million years of its history, the Earth must have been particularly inhospitable. There were numerous major impacts far bigger than that which may have killed the dinosaurs. The scars of this late heavy bombardment can still be seen in the great Maria basins on the Moon, which are themselves giant impact craters filled with basalt lava melted by the impacts. Such impacts would have melted much of the Earth's surface and certainly vaporized any primitive oceans. It is possible that the water on our planet today came from a subsequent rain of comets as well as from volcanic gases.

Dawn of life

The early atmosphere of Earth was once thought to have been a mixture of gases such as methane, ammonia, water, and hydrogen, a potential source of carbon to primitive life forms. But it is now believed that strong ultraviolet radiation from the young Sun would have broken that down quickly to give an atmosphere of carbon dioxide and nitrogen. No one yet knows for certain how life began. There are even claims that it may have had an extra-terrestrial origin, arriving on Earth in meteorites from Mars or beyond. But laboratory studies are beginning to show how some chemical systems can begin to self-organize and catalyse their own reproduction. Analysis of present-day life forms suggests that the most primitive are not the sort of bacteria that scavenge organic carbon or that use sunlight to help them photosynthesize but those that use chemical energy of the sort that is found today in deep-sea hydrothermal vents.

By 3,500 million years ago, there were almost certainly microscopic cyanobacteria and arguably primitive algae – the sort of thing we see today in pond scum. These began to have a dramatic effect. Using sunlight to power photosynthesis, they took in carbon dioxide from the atmosphere, effectively eating the blanket that, by the greenhouse effect, kept the Earth warm when the power of the Sun was weak. This may be what led ultimately to the late Pre-Cambrian glaciation. But long before that, it resulted in the worst pollution incident the world has known. Photosynthesis released a gas that had not existed on Earth before and which was probably toxic to many life forms: oxygen. At first, it did not last long in the atmosphere but quickly reacted with dissolved iron in sea water, resulting in thick deposits of banded iron oxide. Almost literally, the world went rusty. But photosynthesis continued, and free oxygen began to build up in the atmosphere from about 2,400 million years ago, paving the way for animal life that could breathe the oxygen and eat the plants.

Birth of Earth

About 4,500 million years ago there was a great cloud of gas and dust, the product of several previous generations of stars. It began to contract under gravity, perhaps boosted by the shock waves from a nearby exploding star or supernova. A slight rotation in the cloud accelerated as it contracted and spread the dust out into a flattened disc around the proto-star. Eventually, the central mass, mostly of hydrogen and helium, contracted sufficiently to trigger nuclear fusion reactions at its core and the Sun began to shine. A wind of charged particles began to blow outwards, clearing some of the surrounding dust. In the inner part of the nebula, or disc, only refectory silicates remained. Further out, the hydrogen and helium accumulated to form the giant gas planets Saturn and Jupiter. Volatile ices such as water, methane, and nitrogen were driven still further out and formed the outer planets, Kuiper belt objects, and comets.

The inner planets – Mercury, Venus, the Earth, and Mars – formed by a process known as accretion which began as particles bumped into one another, sometimes splitting, occasionally joining together. Eventually, the larger lumps developed sufficient gravitational attraction to pull others to them. As the mass increased, so did the energy of the impacts, melting the rocks so that they began to separate out, with the densest, iron-rich minerals sinking to form a core. The new Earth was hot, probably at least partially molten, from the impacts, from the energy released by its gravitational contraction, and from the decay of radioactive isotopes. It is likely that many radioactive elements in the pre-solar nebula had been created not long before in supernovae explosions and would still have been radioactively hot. So it is hard to see how there could have been liquid water on the surface initially, and it is possible that the first atmosphere was mostly stripped away by the force of the solar wind.

A chip off the block

The formation of the Moon had long been a mystery to science. Its composition, orbit, and rotation didn't fit with the idea that it had split off from the young Earth, formed alongside it, or been captured whilst passing it. But one theory does now make sense and has been convincingly simulated in computer models. It involves a proto-planet about the size of Mars crashing into the Earth about 50 million years after the formation of the solar system. The core of this projectile would have merged with that of the Earth, the force of the impact melting most of the Earth's interior. Much of the outer layers of the impactor, together with some terrestrial material, would have vaporized and been flung into space. A lot of that collected in orbit and accreted to form the Moon. This cataclysmic event gave us a companion which seems to have a stabilizing effect on the Earth, preventing its rotation axis swinging chaotically and thus making our planet a more amenable home to life.

Chapter 3
Deep Earth

The surface of the Earth is covered by a relatively thin, cold, hard crust. Beneath the oceans it is about 7 or 8 kilometres thick; in the continents, 30 to 60 kilometres thick. At its base lies the Mohorovicic discontinuity or Moho, a layer which reflects seismic waves, probably as a result of a change in composition to the dense rocks of the mantle beneath. The lithosphere, the complete slab of cold, hard material on the Earth's surface, includes not only the crust but the top of the mantle as well. In total, the continental lithosphere may be 250 or even 300 kilometres thick. It thins under the oceans, as you approach the mid-ocean ridge system, down to little more than the 7-kilometre crust. The lithosphere is not a single rigid layer, however. It is split into a series of slabs called tectonic plates. They are our principal clue as to how the deep Earth works. To understand what's going on, we must probe beneath the crust.

Digging deep

Only 30 kilometres away from us lies a place we can never visit. If the distance was horizontal it would just be an easy bus ride away, but the same distance beneath our feet is a place of almost unimaginable heat and pressure. No mine can tunnel that deep. A proposal in the 1960s to use ocean-drilling techniques from the oil exploration industry to drill right through the ocean crust into the mantle, the so-called Moho project, was ruled out on grounds of cost and difficulty. Attempts at deep drilling on land on Russia's Kola Peninsula and in Germany had to be abandoned after about 11,000 metres. Not only was the rock difficult to drill, but the heat and pressure tended to soften the drill components and squeeze the hole shut again as soon as it was drilled.

Messengers from the deep

There is one way in which we can sample the mantle directly: in the outpourings of deep-rooted volcanoes. Most of the magma that erupts from volcanoes comes from only partial melting of the source material, so basalt, for example, is not a complete sample of mantle rock. It does, however, carry isotopic clues to what lies beneath. For example, basalt from some deep-rooted volcanoes, such as that in Hawaii, contains helium gas with a high ratio of helium 3 to helium 4, as the early solar system is believed to have had. So this is thought to come from a part of the Earth's interior that is still pristine. The helium gets lost in volcanic eruptions and is slowly replaced by helium 4 from radioactive decay. The basalt in ocean ridge volcanoes is depleted in helium 3. This suggests that it is recycled material that lost helium gas in earlier eruptions and does not come from so deep in the mantle.

Violent volcanic eruptions do sometimes carry in their magma more direct samples of mantle rocks. These so-called xenoliths are samples of mantle rock that have not been melted, just carried along in the flow. They are typically dark, dense, greenish rocks such as peridotite, rich in the mineral olivine, a magnesium/iron silicate. Similar rock is sometimes found in the deep cores of mountain ranges which have been thrust up from great depths.

Slow flow

The magnificent medieval stained-glass windows of Canterbury Cathedral can tell us something about the nature of the Earth's mantle. The windows are composed of many small panes of coloured glass in a leaded frame. If you look at the sunlight filtered through the panes, you will notice that some of them are darker at the bottom than at the top. This is because the glass flows. Technically, it's a super-cooled fluid. Over the centuries, gravity has made the panes slowly sag so that the glass is thicker at the bottom. Yet, to the touch, or, heaven forbid, a hammer blow, the glass still behaves as a solid. A key to understanding the Earth's mantle is the realization that the silicate rocks there can flow in the same sort of way, even though they do not melt. In fact, the individual mineral grains are constantly re-forming, giving rise to the motion known as creep. The effect is that the mantle is very, very viscous, like extremely thick, sticky treacle.

A planetary body scan

The clearest clues to the internal structure of the Earth come from seismology. Earthquakes send out seismic shock waves through the planet. Like light being refracted by a lens or reflected by mirrors, seismic waves travel through the Earth and reflect off different layers within it. Seismic waves travel at different velocities depending on how hot or soft the rock is. The hotter and therefore softer the rock, the slower the wave travels. There are two main types of seismic wave, primary, or P, waves, which are the faster and thus the first to arrive at a seismograph, and secondary, or S, waves. P waves are pressure waves with a push-pull motion; S waves are shear waves and cannot travel through liquids. It was by studying S waves that the molten outer core of the Earth was first revealed. Detecting these seismic waves on a single instrument would not tell you much, but today there are networks of hundreds of sensitive seismometers, strung out around the planet. And every day there are many small earthquakes to generate signals. The result is a bit like a body scanner in a hospital, in which the patient is surrounded by X-ray sources and sensors and computers use the results to build up a 3D image of her internal organs. The hospital version is known as a CAT scan, standing for Computer Assisted Tomography. Its whole-Earth equivalent is called seismic tomography.

The global network of seismographs is best at seeing things on a global scale. It will reveal the overall layering in the mantle and changes in seismic velocities due to high or low temperature on scales of hundreds of kilometres. There are also more closely spaced arrays, originally set up to detect underground nuclear tests, and they, together with new arrays being deployed by geophysicists in geologically interesting regions, have the potential to see structure deep in the mantle on a scale of a few kilometres. And it seems that there is structure on every scale. The clearest things in these whole-Earth body scans are the layers. Below 2,890 kilometres, the depth of the liquid outer core of our planet, S waves will not pass. But several features stand out within the mantle. There is, as we've mentioned, the Mohorovicic discontinuity at the base of the crust, and another at the base of the hard lithosphere. The asthenosphere beneath is softer so the seismic velocities are slower. There are clear layers 410 kilometres down and 660 kilometres down, with another less distinct layer after about 520 kilometres. At the base of the mantle is another, probably discontinuous, layer called the

D" or D double prime layer, which varies from nothing to about 250 kilometres thick.

Seismic tomography also reveals more subtle features. Essentially, colder rock is harder so seismic waves travel through it more quickly than they do through hotter, softer rock. Where old, cold ocean crust dives beneath a continent or into an ocean trench, reflections from the descending slab reveal its passage down into the mantle. Where Earth's hot core bakes the underside of the mantle, it appears to soften and rise in a huge plume.

The mantle is full of mysteries which at first sight seem to be contradictory. It is solid yet it can flow. It's made up of silicate rock which is a good insulator, yet somehow about 44 terawatts of heat finds its way to the surface. It's hard to see how that heat flow could happen through conduction alone, and yet, if there was convection, the mantle would be mixed, so how could it show a layered structure? And how could ocean volcanoes erupt magma with a different mix of tracer isotopes than that believed to exist in the bulk of the mantle, unless there are unmixed regions or layers? Resolving these mysteries has been one of the prime areas of geophysics in recent years.

A diamond window on the mantle

Some of the best clues have come from understanding the nature of the rocks down there. To find what the rocks are like deep within the Earth, you have to replicate the fantastic pressures down there. Amazingly, that's possible with just your finger and thumb. The trick is to get hold of two good, gem-quality diamonds, cut in what jewellers term 'brilliant' cuts, with a tiny, perfectly flat face at the apex of each. Mount them face to face with a microscopic rock sample between the two, and turn a little thumbscrew to force the faces tighter together. The force gets so concentrated between the tiny diamond anvils that it's possible to create pressures more than 3 million times atmospheric pressure (300 gigapascals), just by turning the screw. Because the diamonds are conveniently transparent, the sample can be heated by shining a laser in, and viewed with a microscope and other instruments. This can literally be a window on what rocks are like deep in the mantle.

Professor Bill Bassett was studying a tiny crystal in a diamond anvil one day in his lab at Cornell University. Nothing much had happened when he'd increased the pressure, so he decided to go for lunch. As he was leaving, he heard a sudden 'crack' from the anvil. Certain that one of his precious diamonds had broken, he rushed back and looked down the microscope. The gems were OK, but the sample had suddenly transformed into a new, high-pressure crystal form. It was what is known as a phase change: the composition remains the same but the structure changes, in this case into a more dense crystal lattice.

We know from the composition of xenoliths that at least the upper mantle is made of rocks such as peridotite, rich in the magnesium and iron silicate mineral olivine. Put a tiny sample of this between diamond anvils and turn up the pressure and it goes through a whole series of phase changes. At a pressure of about 14 gigapascals, equivalent to a depth in the mantle of about 410 kilometres, olivine transforms into a new structure called wadsleyite. At 18 gigapascals, 520 kilometres down, it changes again, adopting the structure of ringwoodite, a form of the mineral spinel. That then changes at 23 gigapascals, corresponding to 660 kilometres down, into two minerals, perovskite and a magnesium iron oxide mineral called magnesiowüstite. You'll notice that the phase changes happen at precisely the depths at which seismic waves can be reflected. So perhaps the layers indicate a change in crystal structure rather than composition.

A double boiler?

The 660-kilometre layer, the division between the upper and lower mantle, is a particularly strong feature and the focus of vigorous debate between those who think that the entire mantle is circulating in a huge convection system and those who think that it is more like a double boiler with separate circulating cells in the upper and lower mantle and little or no exchange of material between them. Historically, geochemists tend to favour the double structure as it allows for chemical differences between the layers, whereas geophysicists prefer whole-mantle convection. Present indications are that both might be right, in a compromise solution in which whole-mantle circulation is possible but difficult. Data from seismic tomography would seem at first to favour the double boiler idea. The seismic scans reveal where slabs of subducted ocean crust sink down towards the 660 km anomaly. But they do not seem to pass through it. Rather, the material spreads out and seems to collect at that depth, for hundreds of millions of years. But further scans show where it can break through like an avalanche and continue on through the lower mantle almost to the top of the core.

In June 1994, Bolivia was shaken by a powerful earthquake. It did little damage because its focus was so deep – about 640 kilometres. But at that depth, the rocks should be too soft to fracture. This is a region where a slab of old ocean crust from the Pacific is sinking down beneath the Andes. What must have happened is that a whole layer of rock underwent a catastrophic phase change into the denser perovskite structure. That seems to be necessary before it can sink down into the lower mantle. The explanation solves the mysteries of mantle layering and deep earthquakes at one go.

But there is much that still needs explaining. For example, the slab of ocean crust that is subducting below the Tonga trench in the Pacific is passing into the mantle at about 250 millimetres per year, far too fast for its temperature to even out. Material would reach the base of the upper mantle in just 3 million years and its low temperature should be obvious if it pools there or extends into the lower mantle. But there is no evidence for such a slab. One theory is that not all of the olivine converts into higher-density minerals,

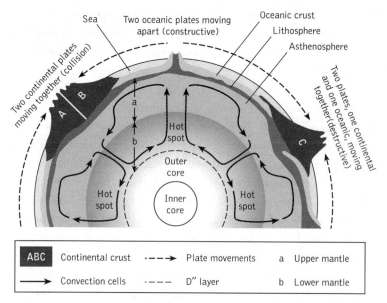

6. The basic circulation in the Earth's mantle and how it is reflected in lithospheric plate motions and plate boundaries. For clarity, motions are simplified and the vertical scale of the lithosphere is greatly exaggerated.

making the old slab neutrally buoyant in the upper mantle. The combination of cool temperature and mineral content would give it a seismic velocity very similar to other mantle material, so it would not show up easily, just as a layer of glycerine does not show up well in water. There is indeed tantalizing faint seismic evidence for such a slab deep below Fiji.

Message in a diamond

Diamond is the high-pressure form of carbon. It can only form in the Earth at depths of over 100 kilometres, sometimes well over this. Isotope ratios in diamonds suggest that they often form from carbon in subducted ocean crust, maybe carbonate from ocean sediments. Sometimes, there are tiny inclusions of other minerals within a diamond. It is not a feature that is popular among gem stone dealers, but it is just what geochemists are searching for.

Minute analysis of those inclusions can tell the long and sometimes tortuous history of the diamond's formation and passage through the mantle.

Some of the inclusions are of a mineral called enstatite, which is a form of magnesium silicate. Some researchers believe that it was originally magnesium silicate perovskite and comes from the lower mantle. Their evidence comes from the observation that it contains only one-tenth as much nickel as would be expected in the upper mantle. At lower mantle temperatures and pressures, nickel gets taken up into a mineral called ferropericlase, which is also a common inclusion of diamonds, leaving very little nickel left in magnesium silicate perovskite. In a few cases, the inclusions are rich in aluminium which, under upper mantle conditions, is locked up in garnet. And some inclusions are iron-rich, suggesting that they might have originated very deep in the mantle, close to the core mantle boundary. These deep diamonds also have a different carbon isotope signature, believed to be characteristic of deep mantle rock rather than subducted ocean lithosphere. Estimates of the age of diamonds and the rock that surrounds them suggests that some have had a very long and tortuous passage through the mantle that may have taken them more than a billion years. But it is convincing evidence of at least some transfer between the lower and upper mantle.

Almost as fascinating as the diamonds themselves is the rock in which they are found. It's called kimberlite after the South African diamond-mining town of Kimberley. The rock itself is a mess! Apart from the diamonds, it contains a whole range of angular lumps and pulverized fragments of different rocks; a so-called breccia. It is volcanic and tends to form the carrot-shaped plugs of ancient volcanic vents. It is hard to determine its exact composition because it contains so much pulverized debris from its passage through the lithosphere, but the original magma must have been mostly olivine from the mantle together with an unusual amount of volatile material now in the form of mica. If it had found its way up slowly from the mantle, we would not have diamonds today. Diamond is unstable at pressures found less than 100 kilometres underground and, given time, would dissolve in the magma. But kimberlite volcanoes did not keep it waiting. It is estimated that the average speed of material through the lithosphere was about 70 kilometres per hour. The widening neck of the vent as it approaches the surface suggests that volatile material was expanding explosively and the surface eruption speed could have been supersonic. As a result, all the rock fragments collected on the way up have been quenched, frozen in time, so that they represent samples from deep in the lithosphere and even the mantle.

The base of the mantle

Recent analysis of seismic data from around the world has revealed a thin layer at the base of the mantle, the D″ layer, up to 200 kilometres thick. It is not a continuous layer but seems more like a series of slabs, a bit like continents on the underside of the mantle. This could be regions where silicate rocks in the mantle are partly mixed with iron-rich material from the core. But another explanation is that this is where ancient ocean lithosphere comes to rest. After its descent through the mantle, the slab is still cold and dense so it spreads out at the base of the mantle and is slowly heated by the core until, perhaps a billion years later, it rises again in a mantle plume to form new ocean crust.

Clues to the deep interior of the Earth also come from measuring tiny variations in day length. Our spinning planet is gradually slowing down due to the pull of the moon on the tides and to the rising of land compressed by ice in the last Ice Age. But there are other even smaller variations of a few billionths of a second. Some may be due to atmospheric circulation blowing on mountain ranges like wind on a sail. But there is another component which seems to be caused by circulation in the outer core pushing on ridges in the base of the mantle like ocean currents pushing on the keel of a ship. So there may be ridges and valleys like upside-down mountain ranges on the base of the mantle. There seems to be a great depression in the core beneath the Philippines that is 10 kilometres deep, twice the depth of the Grand Canyon. Bulging up beneath the Gulf of Alaska is a high spot on the core; a liquid mountain taller than Everest. Maybe sinking cold material indents the core, while hotspots bulge up.

Super plumes

Although much hotter, the perovskite rock of the lower mantle is much more viscous than upper mantle rocks. Estimates suggest that it is 30 times more resistant to flow. As a result, material rises from the base of the mantle in a much slower, broader column than the plumes which characterize the upper mantle. It behaves, in very, very slow motion, rather like the blobs of gloop in a lava lamp. It may well be true that, although some material circulates through the entire mantle, there are also smaller convection cells that are confined to the upper mantle. Convection cells in experimental systems tend to be about the same width as they are deep and, in some parts of the world at least, the spacing of plumes of mantle material seem to match the 660-kilometre depth of the upper mantle.

How the Earth melts

What goes down must come back up again. As plumes of hot mantle rock slowly rise towards the crust, the pressure on them drops and they begin to melt. Scientists can recreate what happens using great hydraulic presses to squeeze samples of artificial rock, heated in furnaces. It's not the entire rock that melts, only a few per cent; producing magma that is less dense than the rest of the mantle and so is able to rise up rapidly to the surface and erupt as basalt lava. How it flows through the remaining rock was another great mystery. It turns out to be down to the microscopic structure of the rock. If the angles at the corners of the little pockets of melt that form between rock grains were large, the rock would be like a Swiss cheese; the pockets would not interconnect and the melt couldn't flow out. But those angles are small and the rock is like a sponge, with all the pockets interconnecting. Squeeze the sponge and the liquid flows. Squeeze the mantle and the magma erupts.

Free-fall

When he saw an apple fall, Isaac Newton realized that the force of gravity was pulling objects towards the centre of the Earth. What he did not know was that apples fall slightly faster in some parts of the world than others – not that it is a difference you normally notice, nor could you easily measure it with apples. But you can with spacecraft. The secret of flying, according to Douglas Adams' *Hitchhiker's Guide to the Galaxy*, is to fall but forget to hit the ground. That's roughly what a satellite does. It's falling freely, but its speed keeps it in orbit. The stronger gravitational pull of a region of dense rock will make satellites speed up. Over a region of lower gravity, they will slow down. By tracking the orbits of low satellites, geologists can build up gravitational maps of the Earth beneath.

When geophysicists compared gravity maps of the surface of the Earth with seismic tomography scans of its interior, they had a surprise. You might expect that cold, dense slabs of ocean crust would result in an excessive gravitational pull because of their higher density, whilst a plume of hot mantle rock rising upwards would be less dense and cause a gravity low. That reality is the opposite way around. The effect is especially pronounced over southern Africa, where a huge plume of hot mantle appears to be rising, and around Indonesia, where cold slabs are sinking. Brad Hager of the Massachusetts Institute of Technology came up with an explanation. The mantle super-plume under southern Africa is causing a huge part of the continent to rise up, higher than you would expect were it simply floating on a static mantle. Southern Africa, he estimates, is elevated by about 1,000 metres above where it would naturally float on the mantle, and this excess uplift of rock causes the gravity high. Similarly, the subducting ocean lithosphere beneath Indonesia is dragging the surrounding surface down behind it, creating a gravity low and resulting in a general rise of sea level compared to the land. Clement Chase, now at the University of Arizona, realized that other broad gravity anomalies corresponded to the ghosts of past subduction. A long band of low gravity that passes from Hudson Bay in Canada, over the North Pole, through Siberia and India, and on to the Antarctic seems to mark a series of subduction zones where ancient sea floor has plunged back into the mantle over the last 125 million years. What was thought to be a rise in sea level which submerged most of the eastern half of Australia about 90 million years ago may have been caused by the continent drifting over an ancient subduction zone that tugged at the region as it passed over, lowering land by more than 600 metres.

The core

We have no direct experience or samples of the Earth's core. But we do know from seismic waves that the outer part of it is liquid and only the inner core is solid. We also know that the core has a much higher density than the mantle. The only material that is dense enough and sufficiently abundant in the solar system to make up the bulk of the core is iron. Although we do not have samples of the Earth's core, we do have pieces of something that's likely to be similar, in iron meteorites. Though not as common as stony meteorites, they are easier to spot. They are believed to come from large asteroids in which an iron core separated out before they were smashed by bombardment early in the history of the solar system. They are mostly made of iron metal but contain between 7% and 15% of nickel. Often, they have a structure of intergrown crystals of two alloys, one containing 5% nickel, the other about 40% nickel, in proportions that give the bulk composition.

An iron core must have formed in the Earth by gravitational separation from the silicate mantle when the new Earth was at least partially molten. As the layers separated, so-called siderophile elements such as nickel, sulphur, tungsten, platinum, and gold that are soluble in molten iron would have separated with them. Lithophile elements would have been held back by the silicate mantle. Radioactive elements such as uranium and hafnium are lithophile, whereas their decay products, or daughters, are isotopes of lead and tungsten so would have been separated out into the core at its formation. That consequently reset the radioactive clock in the mantle at the time the core formed. Estimates of the age of mantle rock put that separation at 4.5 billion years ago, about 50 to 100 million years after the ages of the oldest meteorites which seem to date from the formation of the solar system as a whole.

The inner core

The centre of the Earth is frozen. Frozen at least from the viewpoint of molten iron at the incredible pressures down there. As the planet cools, solid iron crystallizes out from the molten core. Present understanding of the electrical dynamo that generates the Earth's magnetic field requires a solid iron core, but the planet may not have had one for its entire history. There is evidence of the Earth's past magnetic field locked into rocks throughout the Phanerozoic. But most Pre-Cambrian rocks have been so altered that it is difficult to measure any original magnetism. So the only estimate of the age of the inner core comes from models of thermal evolution of the core as the Earth slowly cools. It's the same sort of calculation that Lord Kelvin performed in the late 19th century to estimate the age of the Earth from its rate of cooling. But now we know there is additional heat from radioactive decay. The latest analysis suggests that the inner core began solidifying somewhere between 2.5 and 1 billion years ago, depending on its radioactive content. That may seem a long time, but it implies that for billions of years of its early history, the Earth was without an inner core and perhaps without a magnetic field.

Today, the inner core is about 2,440 kilometres across, 1,000 kilometres smaller than the Moon. But it is still growing. The iron is crystallizing at a rate of about 800 tonnes a second. That releases a considerable amount of latent heat, which passes through the liquid outer core, contributing to the churning of the fluid within it. As the iron or iron-nickel alloy crystallizes out, impurities within the melt, mostly dissolved silicates, separate out. This material is less dense than the molten outer core, so it rises through it in a steady rain of perhaps sand-like particles. It probably accumulates on the base of the mantle like a sort of upside-down sedimentation, collecting in upside-down valleys and depressions. There are seismic hints of a very low velocity layer at the base of the mantle that this upward sedimentation could explain. The sandy sediment would trap molten iron just as ocean sediment traps water. By holding iron within it, the layer provides material that can magnetically couple the magnetic field generated in the core with the solid mantle. If some of this material rises in super-plumes to contribute to flood basalts on the surface, it could explain the high concentrations of precious metals such as gold and platinum in such rocks.

Magnetic dynamo

From the surface, the Earth's magnetic field looks as if it could be generated by a large permanent bar magnet in the core. But it is not. It must be a dynamo, with the magnetic field generated by electrical currents in the circulating molten iron of the outer core. Faraday showed that if you have an electrical conductor, any two out of electrical current, magnetic field, and motion will generate the third. That is the principle on which all electrical motors, generators, and dynamos work. But in the case of the Earth, there are no external electrical connections. Somehow both the currents and the field are generated and sustained by the convection currents in the core. This is what is called a self-sustaining dynamo. But it must have needed some sort of kick-start. Perhaps that came from the Sun's magnetic field before the Earth had one of its own.

The magnetic field on the Earth's surface is relatively simple, but the currents in the Earth's core that generate it must be far more complex. Many models have been proposed, some of which, such as

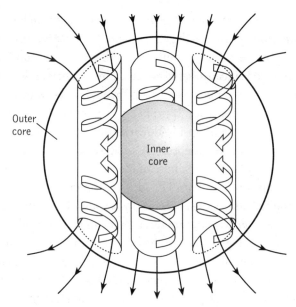

7. A possible model for the generation of the Earth's magnetic field. Convection currents in the outer core spiral due to Coriolis forces (ribbon arrows). That, and electrical currents (not shown) produce the magnetic field lines (black arrows).

the idea of a rotating conducting disc, are purely theoretical. A model that best accounts for the field we see involves a series of cylindrical cells each containing spiral circulation produced by the combination of thermal convection and the Coriolis forces generated by the Earth's rotation. One of the strangest features of the Earth's magnetic field, as we will see in more detail in the next chapter, is that it reverses its polarity at irregular intervals, typically of a few hundred thousand years. At other times there can be periods of up to 50 million years without a reversal. Evidence of the strength of the field trapped in individual volcanic crystals suggests that the field might have been stronger than it is today during such non-reversing periods, or superchrons. The magnetic field is not precisely aligned with the Earth's axis of rotation. At present, it is inclined at about 11 degrees to the Earth's rotation axis. But it hasn't always stayed there. In 1665 it was almost true north, then wandered off, reaching 24 degrees west by 1823. Computer models cannot explain it exactly, but suggest that the dynamo itself is fluctuating chaotically. Most of the time, coupling with the mantle damps out the effects, but sometimes they get so great that the field flips. What is not clear is whether the reversals happen virtually overnight or whether there are thousands of years in which the field wanders wildly or virtually disappears. If the latter is the case, it would be bad news not only for navigation by compass but for life on Earth generally as it would be exposed to more hazardous radiation and particles from space.

There have also been attempts to model what goes on in the Earth's core by experiment. This is not easy since it requires a large volume of electrically conducting fluid circulating with sufficient velocity to excite a magnetic field. It was achieved in Riga by German and Latvian scientists who used 2 cubic metres of molten sodium contained in concentric cylinders. By propelling the sodium down the central cylinder at 15 metres per second, they were able to generate a self-exciting magnetic field.

Taking the Earth's temperature

The deeper you go, the hotter it gets, but how hot is the middle of the Earth? The answer is that, at the boundary between the molten outer core and the solid inner core of the Earth, the temperature must be at exactly the melting point of iron. But the melting point of iron under those incredible pressures will be very different from its value on the surface of the Earth. To find out what it is, scientists must recreate those conditions in their laboratories or calculate them from theory. They've tried two different practical methods: one using tiny samples squeezed between diamond anvils, the other using a giant multi-stage compressed gas gun to compress samples just for an instant. Because of the difficulties in achieving such incredible pressures – 330 gigapascals at the inner core boundary – and because of the difficulty in calibrating pressure so that you know when you've got there, both methods have yet to measure that temperature directly. What they can do is measure the melting point of iron at slightly lower pressures and try to extrapolate downwards. But there are still difficulties. Not least because the core is not pure iron and impurities can affect the melting point. Theoretical calculations put the inner core boundary at about 6,500 degrees Celsius for pure iron, maybe 5,100–5,500 degrees Celsius for iron with the probable range of impurities in the core. These are between estimates from diamond anvil and gas gun experiments.

The study of seismic waves passing through the inner core has produced one more surprise. The waves seem to travel 3–4% faster through the inner core when going north to south compared to east to west; the inner core exhibits anisotropy, a structure or grain which is not the same in all directions. The explanation could be that the inner core is made up of lots of aligned crystals of iron – or even one big crystal more than 2,000 kilometres across! It's also possible that there could be convection within the inner core just as there is in the mantle. And there may be a small amount of liquid caught up in the crystalline mush. It has been calculated that between 3% and 10% by volume of flat discs of liquid aligned with the equator would give the inner core the anisotropy observed.

Spinning core

Like the Earth as a whole, the inner core is rotating, but not exactly in the same way as the rest of the Earth. It is in fact rotating slightly faster than the remainder of the planet, gaining nearly one-tenth of a turn in the past 30 years. Careful study of seismic waves from earthquakes in the South Sandwich Islands off the southern tip of South America that were detected in Alaska show the effect. It is revealed due to the north–south anisotropy in the inner core that we just discussed. As the inner core pulls ahead of the rest of the Earth, the effect due to that anisotropy changes. Seismic waves that skimmed past just outside the inner core arrived in Alaska just as quickly in 1995 as they did in 1967. But waves passing through the inner core made the trip 0.3 seconds faster in 1995 than in 1967, showing that the fast track axis of the inner core has been swinging into alignment at about 1.1 degrees per year. Understanding why the inner core is spinning so fast may give insight into what is going on in that strongly magnetic environment. It could be that currents in the outer core, analogous to the jet streams in the atmosphere, are putting a magnetic tug on the inner core.

So far, only about 4% of the total core has frozen. But, in 3 or 4 billion years' time, the entire core will have solidified and we may lose our magnetic protection.

Chapter 4
Under the sea

Hidden world

Water covers 71% of our planet. Only 1% of that is fresh water; 2% is frozen, and the remaining 97% is salt water in the oceans. It averages more than 4,000 metres deep and reaches down to a maximum of 11,000 metres. All we can easily see is the surface. Scarcely any sunlight penetrates deeper than 50 metres, the so-called photic zone. The rest is a cold, dark world that is alien to us, or at least it was until about 130 years ago.

In 1872, HMS *Challenger* set sail on the first scientific voyage of oceanographic exploration. She visited every ocean and travelled 100,000 kilometres in four years, but depths could only be taken up by single point soundings with a weight lowered over the side of the boat. So the pace of oceanography was slow until the development of techniques such as sonar and sediment coring during the Second World War. During the Cold War, Western powers needed good maps of the sea floor so they could conceal their own submarines, and required advanced sonar and arrays of hydrophones to detect Soviet submarines. Today, ship-mounted and towed sonar scanners have mapped much of the sea floor in considerable detail. The ocean drilling programme has sampled the underlying rocks in many areas, and deep-water manned and robot submersibles have visited many of the most interesting places. But there is still a wealth of exploration ahead.

Where did the water come from?

It seems likely that the Earth's most primitive atmosphere was largely stripped away by the strength of the newborn Sun's solar wind. It is almost certain that the heat generated by the heavy bombardments that completed the formation of the Earth and the huge impact that created the Moon must have melted the surface rocks and driven off most of the original water. So where did our vast oceans come from? There are clues in the oldest rocks, from 4 billion years ago, that liquid water was around when they formed, and there is evidence from not long after that of aquatic bacteria. The oldest fossilized imprints of rain drops are in sediments about 3 billion years old in India. Some of the Earth's surface water may have escaped from the planet's interior in volcanic gases, but most of it probably fell from space. Even today about 30,000 tons of water falls to Earth each year in a fine rain of cometary particles from deep space. In the early history of the solar system that flux must have been significantly higher, and many of the late impacts are likely to have been from whole or fragmentary comets, the composition of which has been likened to dirty snowballs, containing abundant water ice.

Salty seas

Today, about 2.9% by weight of sea water is made up of dissolved salts, mostly common salt, sodium chloride, but also sulphates and bicarbonates and chlorides of magnesium, potassium, and calcium, plus trace elements. The salinity varies depending on the evaporation rate and the influx of fresh water. So, for example, in the Baltic the salinity is low but in the landlocked Dead Sea, it is about six times the average of 35 grams of solids per kilogram of sea water. But the relative proportions of each of the main components of salinity remain constant worldwide.

The oceans were not always that salty. Most of the salt is believed to have come from rocks on land. Some of it was simply dissolved by rain and rivers and some was released by chemical weathering, in which carbon dioxide dissolves in rain to make weak carbonic acid. This slowly converts silicate minerals in rock into clay minerals. These tend to retain potassium but release sodium, which is why sodium chloride is the biggest component of sea salt. For the last few hundred million years, ocean salinity has been approximately constant, with the input of salts from weathering balanced by their deposition in evaporite deposits and other sediments.

The living ocean

The oceans also contain many chemicals in trace quantities and many of these are nutrients important for life and hence for ocean productivity. As a result, they are often depleted in surface waters. Colour scanners flown in space and tuned to the characteristic wavelengths of pigments such as chlorophyll in phytoplankton can map out the seasonal productive zones in the ocean. The highest productivity tends to occur in the spring in mid-to-high latitudes where warm water meets cold but nutrient-rich waters. In the 1980s, the late John Martin of Moss Landing Marine Laboratory in California noticed that blooms of plankton can arise down-current off volcanic ocean islands. He suggested that iron might be a limiting nutrient in ocean productivity and that the volcanic rock was supplying traces of dissolved iron. That has since been confirmed by experiments seeding patches of iron salts in the South Pacific and also by the observation from sediment cores that ocean productivity was highest at the onset of glacial conditions, when wind-blown dust was contributing iron to the ocean. But fertilizing the ocean with iron may not be a cure for the enhanced greenhouse effect as most of the carbon dioxide removed by the plankton seems to get recycled back into solution as they die or are eaten.

Ocean margins

Continents are often fringed by a shallow shelf, little more than 200 metres deep. Geologically, this is effectively part of the continent not the ocean and, at times of much lower sea level, parts of it must have been dry land. The continental shelves are often highly productive and support fisheries, or at least they did until over-fishing started to limit the catch. The organic productivity together with huge quantities of silt, mud, and sand washed by rivers or blown by wind from the neighbouring land has built up thick sediments. Where rivers supply these, the dense, sediment-laden water sometimes continues to flow almost river-like through gorges and over the edge of the continental shelf, sometimes continuing, as in the case of the Amazon, hundreds of kilometres offshore before dispersing in a delta-like pattern. In some places, the shelf margins have spectacular underwater scenery of cliffs and gorges that can only be seen by sonar but are as spectacular as any on land.

The ocean floor

Huge areas of the deep ocean floor are relatively flat and featureless, with little more than the occasional sea cucumber (actually a type of echinoderm, a relative of starfish) for many miles. But there are also mountains and canyons. We'll come to the mid-ocean ridges and trenches later, but there are also many isolated seamounts, sometimes known as guyots, rising from the ocean floor. Literally like underwater mountains, these are often isolated volcanoes, supplied in the past by mantle plumes even though they are not at the margin of a tectonic plate. Many are under more than 1,000 metres of water but carry evidence that they were once volcanic islands that rose above the sea, were eroded flat by waves, and subsided either individually or regionally back to the depths. Sometimes the subsidence was slow enough for coral reefs to build up around the island, leaving, after the volcanic land is gone, a circular atoll. Sometimes there are chains of the islands, formed as the ocean floor moves across a mantle plume. The most famous chain makes up the Hawaiian Islands and the Emperor seamounts to the northwest of Hawaii.

Landslides and tsunamis

The steep margins of the continental shelf and of seamounts means that the slopes can easily become unstable. There is evidence on the sea bed and on surrounding coasts of vast underwater landslides in which slope failure sends many cubic kilometres of sediments cascading down to the abyssal plain. Well-studied examples lie in the Atlantic, west of the islands of Madeira and the Canaries; off the northwest African coast; and off the coast of northern Norway. Sometimes the slide can be triggered by an earthquake, at other times it is simply that sediments have piled up too steeply and the slope fails. Either way, the underwater slides can generate devastating waves called tsunamis. There is evidence of three exceptionally large underwater slides during the past 30,000 years in the Norwegian Sea northwest of Norway. In one, about

7,000 years ago, 1,700 cubic kilometres of debris slid down the continental slope to the abyssal plain east of Iceland. The resulting tsunami flooded parts of the Norwegian and Scottish coastline to a depth 10 metres above the sea level at the time. An even more devastating slide occurred about 105,000 years ago south of the island of Lanai in Hawaii. That island experienced flooding 360 metres above the sea level of the time and the tsunami crossed the Pacific to deposit debris 20 metres above sea level in eastern Australia.

The sediment released by these huge slides and by the more gentle trickle of small slides down continental margins is buoyed up by water in a turbulent flow that can spread it considerable distances. It produces characteristic sediments called turbidites in which the grain size is graded in individual flows. The original slide might contain a mixture of grains but, as the flow fans out, coarse sand falls out more quickly than fine silt and mud, so individual flow bands will have a grading from coarse to fine within them. Such turbidites are often found in sequences of deep-water sedimentary rocks today.

Sea level

One of the clearest features on the surface of our planet is the boundary between land and sea: the coastline. It is one of the most dynamic environments on Earth, with features ranging from high rocky cliffs to low sand dunes and mud flats. And it is an environment into which, for some strange reason, large numbers of humans seem to flock in hot weather. But coastlines do not stay still. Some are eroding as the sea scours away millions of tons of material. In other places land is building up as the sea drives up sand banks or rivers extend their muddy deltas. Over geological timescales, the variations have been spectacular. In some cases huge parts of continents have been flooded in what are called marine transgressions. At other times the sea retreats – marine regressions. These apparent changes in sea level can be due to a number of reasons. One of the present concerns about global warming is that it might cause a rise in sea level. That can be due simply to the oceans warming so the water expands slightly; this alone could raise sea level by maybe half a metre in the next century. It could rise much further if there is significant melting of the Antarctic ice cap. (Melting of the Arctic ice and Antarctic sea ice would have no overall effect on sea level since the ice is already floating and thus already displacing its own weight in water.)

But all this is nothing compared to past sea level changes. Since the peak of the last Ice Age, sea level appears to have risen by as much as 160 metres. It fluctuated dramatically with the climate during ice ages over the last 3 million years. Going back further, sea level was at its highest between about 95 and 67 million years ago, during the upper Cretaceous, when shallow seas covered large areas of continents and thick deposits of chalk were formed together with many of the deposits that are yielding oil today. One theory for these exceptionally high sea levels is that large areas of ocean floor were being uplifted by hot material rising in the mantle as the Atlantic Ocean began to open. The geological record of sea level is characterized by periods of steadily rising seas followed by an apparently abrupt fall in sea level. Sometimes apparent sea level falls can be linked to the tectonic uplift of continents. In some instances it seems to happen globally and not always at the onset of an ice age. Perhaps it is sometimes due to sudden large-scale rifting in the ocean floor, literally pulling the floor out from under the sea.

Drilling the seas

From 1968 the ocean floor was sampled directly and scientifically by the US-led Deep Sea Drilling Program using a drilling vessel named the *Glomar Challenger*. This was superseded in 1985 by the international Ocean Drilling Program using the improved *JOIDES Resolution*. Around 200 separate two-month-long voyages or legs have taken place, with core samples drilled at intervals along each.

The deepest holes exceed 2 kilometres, and overall thousands of kilometres of core samples have been recovered. Many of them include varying depths of sediment down to volcanic basalt beneath. They all tell stories about their origins and the changing conditions of climate and ocean. Far from eroding land and river deltas, sedimentation rates are much lower. At high latitudes the sediments include clay and rock fragments rafted on icebergs which have melted and dropped their load. Elsewhere, wind-blown dust from deserts and volcanic ash makes up a greater proportion of deep-water sediment, sometimes accompanied by micro-meteorite dust, sharks' teeth, and even the ear bones of whales.

8. The ocean drilling ship *JOIDES Resolution*. The derrick towers 60 metres above the water-line.

Where ocean productivity is high on the surface, there are often also the sunken remains of various types of plankton. In comparatively shallow water, the calcareous skeletons of coccolithophores and foraminifera are common, forming a calcareous ooze that may consolidate to form chalk or limestone. But the solubility of calcium carbonate increases with depth and pressure. Somewhere between 3.5 and 4.5 kilometres down in the water, we reach the so-called carbon compensation depth (CCD), below which the tiny skeletons will tend to dissolve away. Here, their place can be taken by siliceous ooze, made from the tiny silica skeletons of diatoms and radiolarians. Silica too would be soluble, but enough gets through to form significant deposits in the Southern Ocean and parts of the Indian and Pacific oceans. In a few places, usually where ocean circulation is restricted, such as in the Black Sea, the bottom water is anaerobic and black shales are deposited. They are sometimes rich in organic material that is not oxidized or consumed in the anaerobic conditions and can slowly turn into oil. Occasionally, the anaerobic deposits are more widespread, reflecting so-called anoxic events, where changes in ocean circulation prevent oxygen-rich waters sinking to the ocean floor.

Messages in the mud

Sediment cores can carry a long and continuous record of past climate. The types of sediments can reveal what was going on on the surrounding land – for instance, material rafted on icebergs or blown from deserts. But more precise records are kept by the ratios of stable isotopes of oxygen in calcareous ooze. Oxygen in water molecules exists in different stable isotopes, principally ^{16}O and ^{18}O. As sea water evaporates, molecules containing the lighter ^{16}O evaporate slightly more easily, leaving the sea water enriched in ^{18}O. That is soon diluted again by rainfall and rivers, except when large amounts of water get locked up in polar ice caps. Then the carbonate taken up by plankton and deposited in sediments will contain more ^{18}O than during the interglacial periods, so the oxygen isotopes in sediments reflect the global climate. By matching up the changes recorded in sediments for a total of more than 20 million years, the ocean drilling programme has shown how the climate fluctuates on timescales that seem to reflect the Milankovich cycle, the wobble of the Earth's axis, and the eccentricity of our orbit around the Sun.

In the 1970s the ocean drilling programme came to the Mediterranean. There, the drill cores reveal something sensational. I was shown one of them where it is now stored at the Lamont Doherty Geological Observatory of Columbia University in New York. It consists of layer after layer of white crystalline material, a mixture of salt (sodium chloride) and anhydrite (calcium sulphate). These evaporite layers can only have been formed by the Mediterranean drying up. Even today, evaporation rates are so high that, were the Straits of Gibraltar sealed off, the entire Mediterranean would evaporate in about 1,000 years. The implication of the hundreds of metres of evaporite in the drill cores are that this must have happened perhaps 40 times between 5 and 6.5 million years ago. When the scientists drilled close to the Straits of Gibraltar, they encountered a chaotic mixture of boulders and debris. This must have been the giant plunge pool of the world's greatest waterfall, when the Atlantic broke through past Gibraltar to refill the Mediterranean. We can only imagine the roar, the spray, the power of the water.

One of the most interesting of the recent legs of the ocean drilling programme involved drilling into deposits of gas hydrate. These are sediments containing high concentrations of methane ice, held in solid form by the low temperatures and high pressures of the deep ocean floor. There is added excitement when a gas hydrate core is returned to the surface as they can easily turn into gas again, sometimes explosively. That has made studying them somewhat difficult, but there are believed to be vast deposits of them. It is possible that they could become an economically important source of natural gas in the future. There are suggestions that they played a significant role in sudden climate change in the past. They can be quite unstable, and an earthquake can make large quantities float free of the ocean floor to rise up in great gas bubbles to the surface. A sudden fall in sea level can also destabilize gas hydrates so that they release their methane, which is a powerful greenhouse gas. It is probable that a sudden global warming 55 million years ago was caused by methane released from gas hydrates. It has even been suggested that some accounts of ships lost in the imaginary Bermuda Triangle in recent times originated from descriptions of large gas bubbles breaking the surface, capsizing boats or asphyxiating their crew.

Large quantities of organic material can become buried in ocean sediments and can, in the right circumstances, get turned into oil. That tends to happen in shallow marine basins that are undergoing crustal stretching. This thins the crust, deepening the basin so that it fills up with more sediment. But at the same time the organic material gets buried deeper, closer to the internal heat of the mantle so that it is cooked into crude oil and natural gas. This can then rise up through permeable strata and collect beneath impervious clays or salt layers. Rock salt is particularly mobile as it is not very dense and tends to rise up through strata in big domes. Often these can trap rich oil and gas deposits, as happens in the Gulf of Mexico.

Life underground

But not all the organic material in ocean sediments is dead. Living bacteria are often abundant in sediments more than 1,000 metres under the sea floor, in rocks over a hundred million years old. It seems likely that they were living in the sea floor mud and remained as it was buried deeper and deeper all that time ago. They don't exactly lead exciting lives but they're certainly not dead. It is estimated that they may divide only once every

1,000 years and live by anaerobically digesting organic material and releasing methane. Some bacteria can also survive at high temperatures, possibly up to the 100 to 150 degrees Celsius at which oil forms, and they may play a significant part in this process. It is possible that 90% of all terrestrial bacteria live underground and together comprise as much as 20% of the total biomass on Earth.

The longest mountain chain on Earth

If you were to drain the water from the world's oceans and reveal the spectacular landscape down there, the biggest feature would not be the great ocean island mountains taller than Everest or the great chasms that dwarf the Grand Canyon, it would be a mountain chain 70,000 kilometres in length: the mid-ocean ridge system. The ridges run around the planet like the seam in a tennis ball. Peppered along their length are volcanic fissures. Sometimes these erupt slowly underwater, producing pillow-shaped clumps of dense black basalt lava, like toothpaste from a tube. These are the zones of creation where new ocean crust is forming as the sea floor spreads.

The North Atlantic Ridge was discovered in the mid-19th century by a ship attempting to lay the first transatlantic cable. The ridges are broad, between 1,000 and 4,000 kilometres wide, and rise slowly towards a central line of peaks, typically 2,500 metres above the deep ocean floor but still a further 2,500 metres below the sea surface. The ridge is offset by numerous transform faults perpendicular to its length, displacing the ridge crest by many kilometres. The crest of the ridge often consists of a double line of peaks with a central rift between them. In the first half of the 20th century, proponents of the theory of continental drift such as Arthur Holmes suggested that the ridges might mark places where convection in the mantle brought new crust to the surface, but it was magnetic surveys which finally confirmed one of the most important discoveries in geology: sea floor spreading.

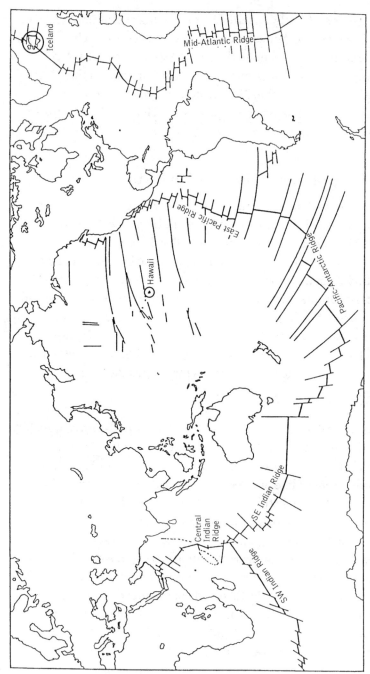

9. The global system of ocean ridges and the main transform fracture zones that cut it. The hot-spots of Hawaii and Iceland are circled.

Magnetic stripes

In the 1950s the US navy needed detailed maps of the ocean floor to aid their submarines. So research vessels began sailing to and fro making sonar measurements. Scientists were given the chance to contribute other experiments, and so it was that a sensitive magnetometer was towed across the oceans, mapping out the magnetic field. The map showed a series of highs and lows in the field strength that appeared like parallel stripes on either side of the mid-ocean ridges. It was Fred Vine and Drum Matthews at Cambridge who were able to confirm what was happening. As volcanic lava erupts and cools, it traps magnetic mineral grains aligned with the Earth's magnetic field. So, sail over recent submarine basalts and the magnetic field of the Earth will be slightly enhanced. But, as we heard in the last chapter, the Earth's

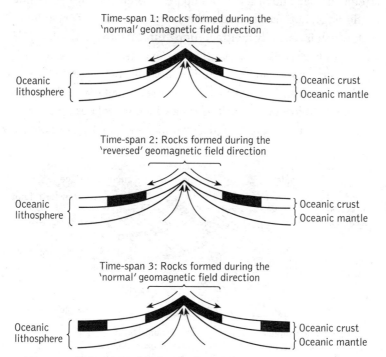

10. How parallel stripes develop in the magnetization of volcanic rocks on the ocean floor, as new ocean crust spreads out from an ocean ridge.

magnetic field sometimes reverses. Magnetism trapped in volcanic rocks that erupted when the field was reversed will carry an opposite component to the present field, lowering the reading slightly. Thus the magnetic stripes on either side of the mid-ocean ridge build up, with older and older sea floor as you move away from the central ridge in either direction. The sea floor is indeed spreading.

Boundary of creation

Overall, the spreading rate is slow but relentless, ranging from about 10 centimetres a year in the Pacific to 3 or 4 centimetres per year in the Atlantic, about the same rate at which your fingernails grow. But the eruption of lava to create new crust is not steady, which is why parts of the ridge get rifted and subside as they are stretched open and others build up peaks. Beneath the centre line of the ridge, hot mantle material is rising in a mush of crystalline rock which is partially melting. Along this line, the hot, soft asthenosphere rises to meet a thin ocean crust with no hard lithospheric mantle in between. Because this mantle material is hot it is less dense and so makes the ridge rise. About 4% of the mantle rock melts to form the basalt magma which percolates up through pores and fissures into a magma chamber a kilometre or so beneath

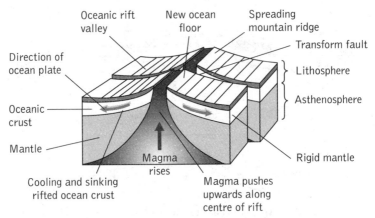

11. **The principal components of a mid-ocean ridge.**

the ridge. Seismic profiles reveal magma chambers several kilometres wide under parts of the Pacific ridge, though they are harder to see under the Atlantic ridge. Material in the magma chamber is slowly cooling, so some crystallizes out and accumulates at the bottom of the chamber to form a coarse-textured rock called gabbro. The remaining melt periodically erupts from fissures along the ridge. It is quite fluid and doesn't contain much gas or steam, so the eruptions are fairly gentle. But the lava is rapidly quenched by the sea water and tends to form into a series of pillow-like structures.

Black smokers

Even where there is not an active eruption, the rocks close to the ridge are still very hot. Sea water gets drawn into cracks and pores in the dry basalt, where it is heated and dissolves minerals such as sulphides. The hot water then rises out of vents, precipitating sulphide to form tall, hollow chimneys. Bacteria that can withstand the hot water contribute to the process by reducing soluble sulphates to sulphides. As the sulphides come out of solution in the cooling water, they form a cloud of black particles, so these vents are often known as black smokers. Water can gush out of them at high velocity and at temperatures in excess of 350 degrees Celsius, making them hazardous but fascinating to explore in deep-diving submersibles. The mineral chimneys can grow at a rate of several centimetres per day until they collapse in a pile of debris. In this way, considerable deposits of potentially valuable sulphide minerals can build up. Where the water is more acidic but slightly cooler, more zinc sulphide is dissolved, creating a white smoker. These are slower growing and generally cooler and thus better habitats for some of the amazing life forms that cluster around such hydrothermal vents. Life down here is based entirely on chemical energy, not sunlight. Primitive bacteria flourish in the hot and often acidic conditions. Blind shrimps, crabs, and giant clams feed off them, and giant tube worms containing symbiotic bacteria filter nutrients from the water. It has been suggested that life on Earth first began in such places, so they are causing much excitement among researchers.

Wealth from the sea

One of the surprise discoveries on the *Challenger* voyages in the 1870s was the return, in dredge samples from the deep ocean floor, of strange black nodules. These nodules are especially rich in manganese and iron oxides and hydroxides, together with potentially valuable metals such as copper, nickel, and cobalt. Known as manganese nodules, they are now known to pepper large areas of the deep ocean floor. Exactly how they form is uncertain, but it seems to be in a slow chemical process with the metals derived from sea water and possibly the underlying sediments. The nodules often grow in concentric onion-like layers around a small, solid nucleus, perhaps a chip of basalt, a bead of clay, or a shark's tooth. Estimates of their age suggest that they are very slow-growing, perhaps adding only a couple of millimetres in a million years. In the 1970s there were various proposals to mine them with scoops or suction, but so far this has not happened due to technological, political, ecological, and economic hurdles.

Pushes, pulls, and plumes

It does not seem as if sea floor spreading is the result of the ocean floor pushing apart from the mid-ocean ridge system. For most of their length, the ridges do not have substantial mantle plumes of hot material rising beneath them. It seems more as if they are being pulled apart, with new material rising to fill the gap. Beneath the ridge there is no thick, hard lithosphere, just a few kilometres of ocean crust. As mantle material rises under the ridge, the pressure drops and so does the melting point of some of the minerals. This leads to partial melting of as much as 20 or 25% of the material, producing basalt lava. The rate of magma formation is just right to produce ocean crust of a fairly uniform thickness of 7 kilometres.

A notable exception is Iceland, where a mantle plume and a mid-ocean ridge coincide. Here, far more basalt is erupted and the crust is around 25 kilometres thick, so that Iceland rises above the Atlantic. The history of that mantle plume can be traced in thickened basalt ocean crust right across the north Atlantic between Greenland and Scotland. Seismic surveys reveal that there are about 10 million cubic kilometres of additional basalt there, several times the volume of the Alps, or enough to cover the entire USA with a layer one kilometre thick. A lot of it didn't erupt on to the surface but was injected beneath the crust, under-plating it. The Hatton bank off the coast of Greenland is a bulge caused by such injections of basalt. The mantle plume that is now under Iceland may have been what caused the north Atlantic to begin to open about 57 million years ago. The volcanic activity appears to have started with a series of volcanoes, some of which are still preserved in the Inner Hebrides and Faeroe Islands northwest of Scotland.

Where oceans go to die

Ocean crust is constantly forming. As a result, it is difficult to find any truly ancient ocean floor. The oldest dates back to the Jurassic, about 200 million years ago, and is in the western Pacific. A segment about 145 million years old was recently discovered near New Zealand. But such ages are rare; most of the ocean floor is less than 100 million years old. So where have all the ancient oceans gone?

The answer lies in a process called subduction. As the Atlantic widens, the Americas on one side and Africa and Europe on the other are slowly moving apart. But the Earth is not getting bigger overall, so something must be taking up the slack. It appears to be the Pacific. The Pacific seems to be ringed by great trenches, up to 11,000 metres deep. Behind them is a ring of volcanoes on islands or on continents, the so-called Pacific ring of fire. Seismic profiling shows how the ocean plate – the thin ocean crust and as much as 100 kilometres of mantle lithosphere beneath – is plunging back

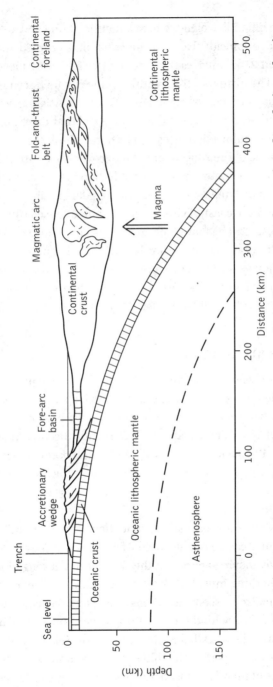

12. How ocean lithosphere subducts beneath a continent, accreting sediments along the margin and producing volcanic activity inland.

down into the Earth. In the 100 million years of its existence, the rocks of the lithosphere have steadily cooled and contracted, becoming more and more dense so that they can no longer float on the asthenosphere. It is this process of subduction that is one of the driving forces of plate tectonics: a pull rather than a push.

The cold, dense rock which sinks in a subduction zone has been under the sea, so it is wet. There is water in the pore spaces and also bound chemically in minerals. As the slab sinks and the pressure and temperature increase, the presence of water lubricates its flow but also lowers the melting point of some of its components, which rise through the surrounding crust to feed the ring of fiery volcanoes. As we saw in the last chapter, the rest of the slab of lithosphere continues down into the mantle, at least to the 670-kilometre boundary with the lower mantle, but eventually sinks perhaps as far as the base of the mantle. Seismic tomography can help trace its billion-year journey.

There are several different types of boundary between the slabs of continent and ocean lithosphere that make up the tectonic plates of the Earth. In the ocean there are the constructive boundaries of ocean ridges and the destructive ones where subduction occurs. This can take place where ocean lithosphere dives down beneath a continent, as in the case of the west coast of South America, forming the volcanic peaks of the Andes. Or ocean can dive beneath ocean, as with the deep trenches of the western Pacific, where the ring of fire comprises volcanic island arcs. There are boundaries where one plate grinds its way alongside another, such as along the coast of California. And there are also plate boundaries where continent runs into continent, but we will discuss them in the next chapter.

What's left on land

Not everything vanishes with a lost ocean. Where ocean lithosphere dives beneath the continents or where an entire ocean is squeezed out between two land masses, much of the sediments get scooped up and added to the continents. That is one of the reasons why so many marine fossils are to be found on land. Occasionally, entire masses of ocean crust can get lifted on to land, a process called obduction. Because they are from collision zones, such rocks are often very distorted, but, by piecing together the evidence from several such sequences, an overall picture emerges. They are known as ophiolite sequences, from the Greek for 'snake rock'. The descriptive name is also reflected in the name serpentinite, given because of the wiggly lines in the green minerals of the forms metamorphosed by hot water. At the top of an ophiolite sequence are the remains of ocean sediments followed by pillow lavas and sheets of basalt that may have been injected underground. Then comes gabbro, the slow-cooled crystalline rock of the same composition as basalt, and at its base the layered deposits of crystals from the bottom of the magma chamber. Beneath that there can be traces of the mantle rock from which the basalt was derived.

Lost oceans

Over hundreds of millions of years it is clear that many oceans have both opened and closed. For a long time, from 1,200 to 750 million years ago, the continents were clustered into one giant super-continent, surrounded by a single vast ocean spanning two-thirds of the globe. In the late Pre-Cambrian, the super-continent broke up into separate land masses. New oceans formed. One of them, the Iapetus Ocean, lasted between about 600 million and 420 million years ago. The join, or suture, where it closed again can today be crossed in a short drive across northwest Scotland. Half a billion years ago that journey would have involved a 5,000-kilometre sea crossing. By the Jurassic period of about 200 million years ago, a great wedge of ocean, the Tethys, had opened up between western Europe and southeast Asia where it opened into the Pacific. That closed as Africa hinged round into Europe to form the Alps and India came crashing into Tibet, lifting the Himalayas. Seismic studies trace remnants of the Tethys ocean floor descending into the mantle.

Over geological time, there have been numerous occasions when new oceans might have formed but did not. East Africa's Great Rift Valley and the Red Sea and Jordan Valley are obvious recent examples. The stretching of the North Sea basin, which produced North Sea oil deposits and Bavarian hot springs, is another. Another few hundred million years and our ocean charts will be completely out of date again.

Chapter 5
Drifting continents

As a child I used to enjoy helping my mother to make marmalade. I confess that I still like making it occasionally myself. But now, when I stare into the preserving pan of simmering fruit and sugar,

I can't help imagining that I am seeing our planet's evolution, greatly speeded up with one second perhaps representing ten or even a hundred million years. When the jam is gently simmering on a slow heat, convection cells establish, with columns of hot marmalade rising to the surface and spreading across it. With them comes some scum, a fine sugary foam, which is not dense enough to sink back down but collects in rafts on the calmer areas of the surface. This foam is a bit like the Earth's continents. It starts to form quite early on in the process and slowly builds up and thickens. Occasionally, the convection pattern beneath changes and the scum splits apart. Sometimes the rafts of scum run together and pile up even thicker. Of course, we shouldn't take this analogy too far. The timescales and the chemistry are altogether different; by and large geologists don't find sugar crystals in granite or orange peel xenoliths in basalt. But it's an image worth holding in mind as we consider the scum of the Earth: the continents.

Scum of the Earth

Continental crust is very different from the crust that floors the oceans. Ocean crust is predominantly magnesium silicate, whereas the continents contain higher proportions of aluminium silicates. They also contain less iron than the denser material of the mantle or of the ocean floor. As a result, they float, albeit on the semi-solid mantle rather than in liquid. And they can be thick. The ocean crust is a fairly uniform 7 kilometres thick, but the continents can range from 30 to 60 kilometres or more. And, like the ocean lithosphere, they are under-plated by a thick layer of cold, hard mantle. Just how deep the roots of continents do go is still a subject of controversy that, in the end, probably comes down to definitions. But continents are also a bit like icebergs: there's a lot more below ground than we can see above. And the higher they rise in mountain ranges, by and large the deeper they go beneath.

Drifting continents

With the benefit of hindsight, the knowledge of mantle convection, and the evidence of sea floor spreading, it is very easy to see that the continents have moved over geological time relative to one another. But it was not always so convincing. In spite of James Hutton's ideas about mountain-building and the rocks cycle, it was a long time before any mechanism could be suggested. Between 1910 and 1915, the American glaciologist Frank Taylor and the German meteorologist Alfred Wegener proposed the hypothesis of continental drift. Yet no one could imagine a way in which the continents could drift like ships at sea through the seemingly solid, rocky mantle. For nearly half a century, supporters of continental drift were in the minority. But the theory's few supporters were working hard. Alex du Toit in South Africa was building up evidence of similar rock structures between southern Africa and South America, while Arthur Holmes, a British geophysicist, proposed mantle convection as a mechanism for the drift. It was not until the 1960s, when the oceanographers got to work, that the debate was settled. Harry Hess proposed that convection beneath the ocean crust might cause the sea floor to spread out from mid-ocean ridges, and Fred Vine and Drum Matthews provided the magnetic evidence of sea floor spreading. It was papers by Tuzo Wilson in Canada, Jason Morgan at Princeton, and Dan McKenzie at Cambridge that brought the evidence together into the theory of plate tectonics.

Plate tectonics explains the surface of the Earth in terms of the motions of a small number of rigid plates which move relative to one another, interacting and deforming along their boundaries. It is not that the continents are drifting free but that they are carried on plates which extend far deeper to include the mantle lithosphere, typically 100 kilometres thick. The plates are not restricted to the continents, but include the slabs of ocean floor as well. There are seven principal plates: the African, Eurasian, North American, South American, Pacific, Indo-Australian, and Antarctic plates.

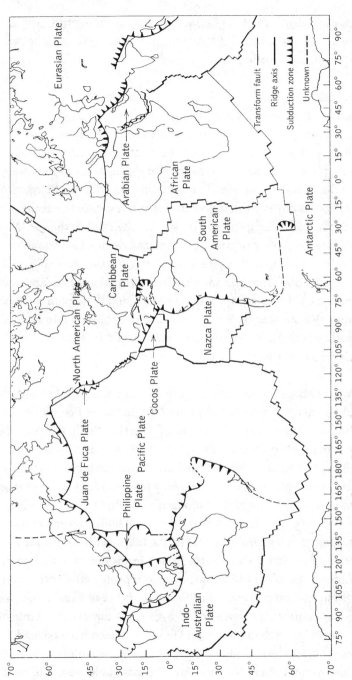

13. The major tectonic plates of the world and their boundaries.

There are also a number of smaller plates, including three quite substantial ones around the Pacific, plus some more complicated fragments where other plates join.

Another of my childhood memories is of tracing the continents from an atlas, cutting them out and trying to fit them together as a single land mass. This must have been about the time of Tuzo Wilson's paper in *Nature* in 1965. I can still remember the thrill of finding how well they fitted and of discovering some of the reasons why the fit was not perfect. It was not just down to my inaccurate tracings. As any nerdish schoolboy knows, you have to cut the continents at the edge of the continental shelf rather than at the coastline. And you can cut off the Amazon delta which would otherwise overlap with Africa, since that has grown since the split of the continents. More exciting was discovering that North and South America need to come apart to make a fit, and that Spain must part company from France. Swinging it back bangs Spain into France exactly where the Pyrenees are today. So could such continental collisions be the cause of mountain ranges?

It was at about that time in my teenage years that family holidays took me to the Pyrenees and to the Alps. In places I could see the layers of sedimentary rock not lying flat as they did in less disturbed lands but rucked up like a carpet into folds and undulations. This took my thoughts back to marmalade. As the conserve simmers, you keep a china plate in the fridge. Every few minutes you bring it out and drip a few drops of the hot marmalade on to it. When it has cooled, you push a finger into it. If it is still liquid, there's nothing for it but to lick your finger and let the marmalade continue to simmer. But after a while, as the brew approaches its setting point, a sample on the plate will crinkle up as you push your finger into it like a miniature continental collision. And it's not a bad model for the way continents behave on large scales. Compressed to fantastic pressures by overlying rocks, and possibly heated from beneath, rocks subjected to the lateral force of a colliding continent will tend to fold rather than to fracture. And the incredible masses of rock involved will be strongly affected by gravity so that the steepest folds will sag under their own weight into over-folds, rather as the skin on custard, or indeed marmalade, would do.

The Earth is not flat

Another reason why flat continents cut from an atlas do not fit together very well is that they are supposed to represent plates on the surface of a sphere. They get distorted in the map projection. But it is not easy to slide rigid plates about on the surface of a sphere either. You cannot simply move them in straight lines because there are no straight lines on a sphere. Each motion is in effect a rotation about an axis cutting through the sphere. But there are still difficulties. One is finding a frame of reference among all the jostling plates. Another is accommodating different rates of sea floor spreading. A simple model might invoke an axis similar to the Earth's rotation axis for the opening of the Atlantic and the relative motions of the Americas away from Africa and Europe. But that would require creating Atlantic Ocean crust like a segment of orange skin, wide at the equator and narrowing smoothly towards the poles. The rate of sea floor spreading does vary, but not in a convenient way like that. The result is transform faults; breaks in the crust thousands of kilometres long, offsetting segments of the mid-ocean ridge.

Frames of reference

With the evidence of sea floor spreading and the mechanism of mantle convection, plate tectonics rapidly became established at the centre of modern Earth sciences. But even today there are geologists who object to the term 'continental drift' because of its associations with the time when the mechanism had not been properly explained and few believed it. But, once people were prepared to accept it, evidence for past plate motions became obvious. There was the geological evidence of rocks of the same type split apart and now lying on opposite sides of an ocean. There was evidence from living and fossil remains of the times different populations became isolated from one another or when they were able to cross between continents. For example, it is just over 200 million years since Australia parted company from parts of Asia such as Malaysia and Indochina. Since then, mammals have been evolving independently on the two land masses, with the result that the marsupial line has come to dominate in Australia while placental mammals developed in Asia.

As with the evidence of magnetic reversals in ocean floor basalts that we discussed in the last chapter, so the magnetic evidence has provided the most comprehensive picture of past continental movements. The grains of magnetic mineral trapped like tiny compass needles in volcanic rocks when they solidified record the direction to the North Pole at the time. They show not only the small wiggles and big reversals in the magnetic field itself, but also trace out over tens or hundreds of millions of years, a larger, more sweeping series of curves: a so-called polar wandering curve. This is in effect the plot of how the continent itself has moved relative to the magnetic pole. When you compare the curves of different continents, sometimes you see that they move together but at other times they diverge, tracking how the continents themselves have split, drifted apart, and come together again in a sort of continental waltz. In fact, it's more like a clumsy barn dance, as the continents occasionally barge into one another.

Sensitive instruments even make it possible to track the relative continental movements today. Over short distances, such as locally across plate boundaries, surveying techniques and in particular laser ranging can be very accurate. But that can now be done over continental scales too, via space. Some of the strangest space satellites ever launched are for laser ranging. The satellite consists of a sphere of dense metal such as titanium with many glass reflectors, like cats' eyes, embedded in it. These reflect light back in the same direction from which it has come, so if you shine a powerful but narrow laser beam from the ground and time the reflected pulse back to where it started, you can work out the distance to an accuracy of centimetres. When you compare values from different continents you start to see how they are moving from one year to the next. Astronomers can do the same sort of thing with radio telescopes using distant cosmic sources of radio waves as their frame of reference. Now that the codes from the American military Global Positioning System (GPS) satellites are no longer scrambled, geologists can get similar precision using a small hand-held GPS receiver in the field. By the careful use of many readings, the accuracy comes down to millimetres. The answers confirm evidence of the rate of sea floor spreading: the plates are moving relative to one another at roughly the speed your fingernails grow, between about 3 and 10 centimetres per year.

Of course, all these plate motion measurements are relative to one another, and it is hard to establish a background frame of reference for them all. A clue comes from Hawaii. The Big Island of Hawaii is just the latest in a series of volcanic ocean islands that stretch away to the northwest and continue underwater as the Emperor seamount chain. Dates of the basalt reveal that the further northwest you go, the older the basalts are. It seems as if the chain marks the passage of the Pacific plate across an underlying plume of hot mantle material. Comparing the historical position of this with those of other mantle plumes shows little relative movement between them, so perhaps these mantle plumes are reference points in an underlying mantle which changes little. Estimates of absolute plate motions, relative to this framework, show that the western Pacific is moving the most. By contrast, the Eurasian plate is scarcely moving at all, so perhaps the historical choice of Greenwich as the reference point for longitude is geologically valid!

The continental waltz

Using a combination of geological and palaeomagnetic evidence, it is possible to trace back the motions of the tectonic plates through geological time. The continents we see today come from the

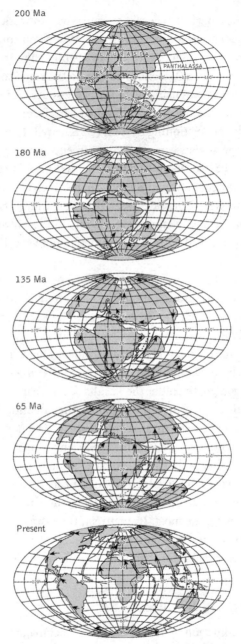

14. The changing map of the Earth's continents over the last 200 million years.

break-up of a super-continent that has been named Pangaea. That split apart in the Permian period about 200 million years ago, initially into a northern continent called Laurasia and a southern continent of Gondwanaland. The break-up of those continents continues today. But going further back in time, it seems that Pangaea itself was made up from an accumulation of earlier continents and that, still further back, there was an even earlier super-continent which has been called Pannotia, and one before that called Rodinia. These cycles of super-continent break-up, drifting, and re-accumulation have been called Wilson cycles, after Tuzo Wilson.

The further back in time you go, into the Pre-Cambrian, the less clear the picture becomes and the harder it is to pick out the land masses we know today. For example, in the Ordovician period, around 450 million years ago, Siberia was near the equator and most of the land masses were gathered in the southern hemisphere, with what is now the Sahara Desert near the South Pole. Late in the Pre-Cambrian, Greenland and Siberia were far south of the equator, Amazonia was almost at the South Pole, whilst Australia was well into the northern hemisphere.

One of the current record-holders for long-distance drifting must be the land known as the Alexander Terrane. It now forms a large portion of the Pan Handle of Alaska. About 500 million years ago, it was part of eastern Australia. The palaeomagnetic evidence in the rocks includes an inclination from the horizontal, dipping down into the Earth, which reveals the latitude at which the rock formed. The steeper the inclination, the higher the latitude. Other clues come from tiny grains of a mineral called zircon. These carry within them products of radioactive decay that date the periods of tectonic activity when they formed. For the Alexander Terrane, they reveal two major mountain-building episodes, 520 and 430 million years ago. Eastern Australia was the site of mountain-building at both these times, whilst North America was quiet. Conversely, western North America was active 350 million years ago, a time when the

Alexander Terrane seems to have been dormant. The Alexander Terrane began to split from Australia 375 million years ago and formed a submarine ocean plateau, at which time various marine animals were fossilized there. About 225 million years ago, the Terrane began moving northward at 10 centimetres a year. This continued for 135 million years, whereupon North American fossils start to appear as the Terrane reached its present latitude and collided with Alaska. It is even possible that it brushed past the coast of California on its way, scraping off material from the California Mother Lode gold belt. If that is correct, the Alaskan gold rush may have been to the same rocks as the Californian gold rush, but displaced 2,400 kilometres to the north.

Continental pile-up

We have heard about several different types of plate boundary. There are the spreading centres of mid-ocean ridges, the transform faults perpendicular to the ridges, and the subduction zones where ocean lithosphere takes a dive under a continent. All are comparatively narrow, well-defined zones that are relatively easy to understand and explain with simple diagrams. But there is one type of plate boundary that is more complex and where the idea of the tectonics of rigid plates breaks down: intercontinental collisions. Where ocean crust is involved it is comparatively easy. As long as it is cold, it will be dense enough to sink down into the mantle at a comparatively steep angle of about 45 degrees. Continental crust won't sink, just as a cork floating in the sea stays afloat in spite of all waves breaking over it. Continental crust also deforms more easily than ocean lithosphere. So, when continents collide, it is more like a serious traffic accident.

A good example is how India joined Asia. For hundreds of millions of years, India had been one of the partners in a complex country dance across the southern hemisphere featuring Africa, Australia, and Antarctica. Then, about 180 million years ago, it broke away and began to drift northwards. There are some spectacular mountain ranges down the western side of India, the Western Ghats and Deccan Traps. A strange feature of them is that, although sea is quite close to the west, the major rivers that drain these ranges flow to the east. Another puzzle came when Professor Vincent Courtilliot of the University of Paris started to look at palaeomagnetism in the basalt rocks of which the hills are made. He had been working in the Himalayas and wanted some comparisons from further south. He expected to find millions of years' worth of palaeomagnetic data in the thick basalt layers, spanning many palaeomagnetic reversals. Instead, he discovered that they were all magnetized in the same direction, suggesting that they must have erupted within one short period of at most a million years. Keith Cox at Oxford University and Dan McKenzie at Cambridge worked out what must have happened. India was once a larger continent, and as it drifted north, it passed over a mantle plume at just the time it was producing a huge pulse of magma. This caused the continent to dome up. But the Deccan Traps are just on the east side of this dome. Rivers could not drain to the west since that lay uphill. Unimaginable volcanic eruptions produced several million cubic kilometres of basalts in the space of thousands of years. Eventually, the activity split the continent in two. The Indian subcontinent we know today is just the northeastern portion of that. The rest lies under the sea in the huge basalt bank between the Seychelles and the Comoros. It turns out that this volcanic outpouring occurred around 65 million years ago, around the time of the Cretaceous/Tertiary boundary, and the extinction of many animal groups including the dinosaurs. Maybe it was not an asteroid after all that killed them but the pollution and climate change that resulted from these incredible volcanic eruptions.

Meanwhile, what remained of the subcontinent continued northwards, closing the great oceanic gulf of the Tethys and eventually ramming into Asia. Whilst the ocean lithosphere of the Tethys was dense and could subduct back into the mantle beneath Asia, the continental crust could not. The two continents first came into contact about 55 million years ago, but a continent has such momentum that nothing is going to stop it dead in its tracks. The closing speed was about 10 centimetres per year. It slowed to around 5 centimetres per year but has continued colliding ever since, like a vehicle crash test played in very slow motion. During this time, the Indian subcontinent has moved a further 2,000 kilometres north. The first thing to happen was a pile-up of sediments and a thickening of the crust in a series of under-thrusts as slabs of Indian continent wedged beneath Asia, like the debris in

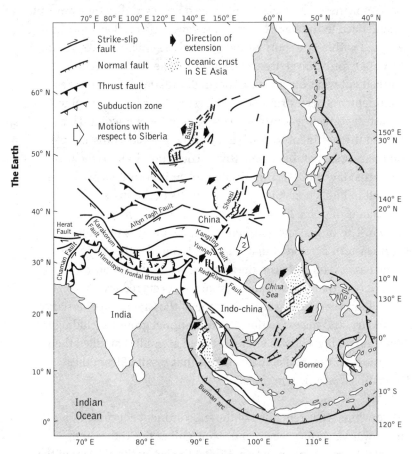

15. Tectonic map of Southeast Asia showing the principal fractures resulting from the collision of the Indian subcontinent and the motions of China and Indochina as they are squeezed to one side.

front of a bulldozer. This thick continental material gave rise to the high Himalayas.

As you head north across the flat Ganges plain, the first of these giant thrust faults forms a sharp feature. Here and there in the line of hills are sediments that were once deposited on the river bed but are now lifted tens of metres above it, yet they are only a few thousand years old, suggesting that there has been sudden and dramatic uplift in earthquakes. These hills are the foothills of the Himalayas which rise in a series of ranges stretched out to the east and west. Each line of mountain peaks corresponds approximately to another huge wedge of continental rock. The rocks exposed in the Himalayas today are mostly ancient granites and metamorphic rocks that have been uplifted from deep within the continents. The folded sediments scooped up from the floor of the Tethys Ocean lie north of the mountains on the edge of the Tibetan plateau. Behind them is a line of lakes corresponding to the original join between the continents, known as the suture.

An old, cold continent such as India is hard and relatively rigid. The part of Asia it collided with is relatively young and soft. Just as the hot mantle can undergo solid flow, so can crustal rock. We think of rock as we find it at the surface, hard and brittle, but crustal minerals such as quartz can flow like toffee at temperatures of only a few hundred degrees Celsius, just as olivine can deeper in the mantle. Some of the best models for the collision of India with Asia come if India is taken to be a relatively solid continent being driven into something with many of the properties of a liquid. And it is a liquid rather like non-drip paint – the harder you push on it, the easier it becomes to deform it. Models of this sort can account for the patterns of mountain ranges in central Asia but not for the high plateau of Tibet.

The rise of Tibet

The pile-up of relatively low-density crustal rocks could not simply be accommodated by downward thrusting, and the entire region began to float upwards. The dense lithospheric root beneath Tibet detached and sank back into the underlying asthenosphere. The remaining thickened continental rocks floated upwards, lifting the Tibetan plateau by as much as 8 kilometres. At the same time, parts of Asia tried to slide out of the way, with Indochina heading eastwards. This sideways motion stretched the continent further north, causing, among other features, lake-filled rifts in Tibet and the deep rift of lake Baikal in Russia. With some of the underlying cold, dense lithosphere removed, the hot asthenosphere was sufficiently close to the Tibetan crust to cause localized melting and account for the recent volcanic rocks found in parts of that country. There is also seismic evidence for a vast pool of partially molten granite about 20 kilometres under the southwestern part of the Tibetan plateau. That would also help to explain how Asia absorbed the impact of India and why the Tibetan plateau has remained relatively flat though surrounded by high mountain ranges. It seems that, overall, the Tibetan plateau is unlikely to get any higher than its present average of 5,000 metres. Any additional uplift would be balanced by a flow of material away to the sides. The more mountainous areas too have an average elevation of no more than about 5,000 metres above sea level. Here, heights are kept in check by erosion. Although a great deal of material has already been eroded from the Himalayas, regions such as Nanga Parbat in the north of Pakistan are still rising today by several millimetres per year, making slopes unstable and prone to landslides.

Monsoon

The Himalayas reach almost as high as passenger aircraft normally fly, and the mountains pose a significant barrier to atmospheric circulation. The result is that central Asia to the north remains cold in winter and dry for most of the year. In the summer, warm air rising above the Tibetan plateau holds back moist air from the southwest, so that the clouds build up and release their moisture in the torrential rain of the Indian monsoon. The monsoon stirs up the Arabian Sea, bringing nutrients to the surface and resulting in an annual plankton bloom. This in turn leaves its trace in the sediments beneath. Sediment cores show that this sequence began around 8 million years ago, perhaps corresponding to the end of the major uplift of the Tibetan plateau and the origin of the monsoon weather pattern. Wind-blown dust in China shows that the region north of the Himalayas was drying out at about this time too. There is also a change in the sediments off the west coast of Africa, with an increase in wind-blown dust in the layers. It seems that this corresponds to the start of the drying out of Africa and the beginnings of the Sahara Desert as the moist clouds were drawn away towards India. There is a theory that the huge amount of chemical weathering that must have taken place in the eroding Himalayas drew down so much carbon dioxide from the atmosphere that, in turn, it may have set the stage for the ice ages of the last 2.5 million years. So perhaps the climatic changes in Africa that provided the evolutionary pressures that led to the development of modern humans there also have their origins in the rise of Tibet and the Himalayas.

Swiss roll

Further west than the continental pile-up of the Himalayas, the Tethys Ocean narrowed to an inlet, but the results of the collision, in this case of Italy and the African plate with Europe, are similar, if on a slightly smaller scale. The Alps are one of the most studied and best understood mountain ranges. To the north lies a sedimentary basin which slowly filled with sediments known as molasse. South of the Alps, in Italy, lies the plain of the River Po, equivalent to the Ganges plain in India. Between it and the mountains is a series of wedges of sediment, scooped up from the Tethys Ocean, sediment known as flysch. Then come the high Alps of Switzerland, made of the crystalline base of the continent, together with intrusions of granite from partial melting beneath. Beyond them come a series of very strongly folded rocks scooped up into giant over-folds called nappes, folding to the north and sagging under their own weight, as if they had been scooped up like whipped cream. These nappe folds are often so extensive that older rocks are folded up above younger rocks in a very confusing sequence. As with the Himalayas, there is a series of thrust faults, in places doubling the thickness of the continental crust.

Cratons

No continent is an island entire of itself. Continents can split apart or join up and merge. Modern mountain ranges such as the Alps and the Himalayas are just the latest examples of this. Others are so ancient that they have been worn down almost flat again. The Caledonian range of northwest Scotland and the Appalachians of North America are examples dating from when a forerunner of the Atlantic closed about 420 million years ago. The modern continents are patchwork quilts of such features. But the older and thicker a continent becomes, the more rigid it grows and the longer it lasts. The most stable cores of continents that are least affected by tectonic movements are known as cratons, and they make up the cores of present-day North and South America, Australia, Russia, Scandinavia, and Africa. Over time, they can often undergo slow subsidence. Lake Eyre in Australia and the Great Lakes of North America occupy such basins. The southern African craton, by contrast, has been uplifted by the buoyant rocks of a mantle plume beneath.

Profile of a continent

The same sort of principle that revealed the structure of the Earth as a whole, seismic tomography, can study the deep interior of continents in great detail. To get the high resolution required, the technique relies not on random natural earthquakes on the other side of the world, detected by widely spaced seismometers, but creates artificial seismic waves and picks up their reflections using nearby, closely spaced arrays of detectors. It is very expensive and at first was the monopoly of the oil exploration companies, which jealously guarded the results. But now there are many national projects which are sharing their data. The most advanced of these are in North America, where the Consortium for Deep Continental Reflection Profiling in the USA and Lithoprobe in Canada have built up a detailed series of profiles. To create the shock waves they employ a small fleet of purpose-built trucks which use hydraulic rams to shake the ground with heavy metal plates. Deep vibrations are monitored by a network of sensors over many miles which record the reflections from numerous layers beneath the ground. Computer analysis reveals each discontinuity or sudden change in

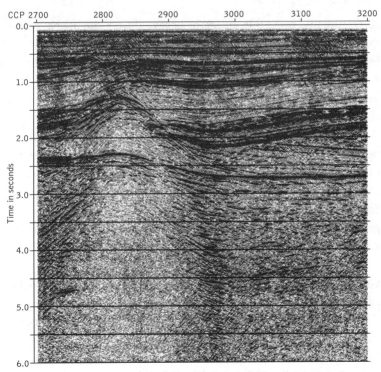

16. Example of a seismic reflection profile of layers and a domed structure within the Earth's crust.

density. These profiles go far deeper than the sedimentary basins of most interest to the oil prospectors. They reveal the ancient sutures between continents that merged with one another long ago. They have revealed reflections from a layer descending into the mantle beneath the Lake Superior region of Canada that could be the oldest subduction zone yet found, with the floor of a lost ocean about 2.7 billion years old. The profiles reveal how basalt magma rising from the mantle and unable to break through the thick continent under-plates it with sheets of basalts known as dykes. They also reveal how, when continental rocks get buried deep enough, they begin to melt so that they rise up through the continent to recrystallize as granite.

The rise of granite

As continental rocks pile up, so the base of the continent gets buried deeper and deeper. As it sinks, it heats up and the rocks at base begin to melt. Many of them are ancient sediments deposited in seas billions of years ago. They contain water chemically bonded into the rocks. The water helps them to melt and lubricates them so that they rise easily towards the surface. Unlike volcanic rocks, they're too thick and sticky to erupt from volcanoes. Instead, huge bubbles of molten rock, perhaps many thousands of metres across, push up into the higher layers of the continents, perhaps at quite high speeds. They bake the surrounding rock and cool slowly, forming a coarsely crystalline rock of quartz, feldspar, and mica: granite. Eventually, the surrounding rock wears away to reveal the great granite domes of, for example, Dartmoor.

Granite may be the inevitable result of a tectonically active planet of silicate rocks and plenty of water. But there can never be a Waterworld, a planet with no continents but global oceans. Once there is water, it finds its way into the chemistry of the rocks, lubricating them as they melt so that they can rise as great masses of granite to form the peaks of continents above the oceans. Without water, you get the situation on Venus: tectonics without the plates.

Without the inner fire of molten magma, you get the situation on Mars, an old, cold surface where life, if it exists, is deep in hiding. On Earth, you get oceans and continents in dynamic and sometimes lethal interaction.

Riches in the earth

One of the first incentives to geological exploration was the search for mineral wealth. Rare and valuable substances can be formed or concentrated by a number of geological processes. Organic remains in sedimentary basins can be gently cooked by the Earth's heat to produce coal, oil, and natural gas. We've already seen how sulphides of valuable metals can be concentrated around deep-sea hydrothermal vents and how manganese nodules can form on the deep ocean floor. Minerals can be concentrated in continental rocks in a number of ways. In molten rock, crystals will begin to form and the densest ones will sink to the bottom of the melt chamber. As well as concentrating minerals within it, a mass of molten rock rising through other rocks will drive super-heated water and steam ahead of it. Under pressure, that can dissolve many minerals, especially those rich in metals, forcing them through cracks and fissures where they are deposited as veins of mineral. Other minerals can become concentrated near the surface when water evaporates or when other components in a rock are eroded away. If we have the technology to recover them, the Earth's riches are there for the taking.

The search for lost continents

If continental scum has been accumulating on the surface of the planet for most of its history, when did it begin? Where is the first continent? It is not easy to say. The most ancient continental rocks have been so reworked, folded, fractured, buried, partially melted, folded and fractured again, and shot through by younger intrusions, that it is hard to make sense of them. It's a bit like trying to identify the remains of an individual car within the compacted scrap from a junk yard. But the search for the oldest rocks on Earth may be nearing its end. Some of the first contenders were from the Barberton greenstone belt in South Africa. These are more than 3.5 billion years old, but they are the remains of pillow lavas and ocean islands, not continents. Similar rocks have now been discovered in the Pilbara region of western Australia, and there are rocks in southwest Greenland that yield dates of 3.75 billion years, but these again are ocean volcanic rocks. The best candidate for the first continent lies in the heart of northern Canada. In the uninhabited, barren lands about 250 kilometres north of Yellowknife, close to the Acasta River, there stands a lonely shed filled with geological hammers and camping equipment. Above the door is a rough sign, 'Acasta City Hall, founded 4 billion years ago'. Some of the rocks from around there have yielded dates a fraction over 4 billion years old.

They have given up their secrets thanks to grains of the mineral zircon, which traps within its lattice uranium atoms, which decay into lead. The grains can be disturbed by re-melting, later growth, and cosmic ray damage, but an instrument developed in Australia known as a SHRIMP (a Sensitive High Resolution Ion Micro Probe) uses a narrow beam of oxygen ions to blast atoms off tiny portions of the zircon so that different zones of the grain can be analysed individually. The centres of some of the grains have given ages of 4.055 billion years, making them the oldest rocks on Earth, and evidence of continents less than 500 million years from the Earth's formation.

Eternity in a grain of sand

But there is tantalizing evidence for something even older. About 800 kilometres north of Perth in western Australia, in the Jack Hills, there are rocks of conglomerate, a mixture of rounded grains and pebbles bound into rock about 3 billion years ago. Among the grains within that rock are zircons that must have eroded out from even earlier rock. One of these has given an age of 4.4 billion years, and an analysis of oxygen atoms in the crystals suggests that the Earth's surface at that time must have been cool enough for liquid water to condense. This research suggests that there were continents far earlier than anyone had expected, within a hundred million years of the Earth's accretion, and seems to run contrary to the concept of a partially molten, inhospitable world at that time.

Super-continents of the future

We have spent most of this chapter looking back in time at the continental waltz of the past. But the continents are still on the move, so what will the world map look like in another 50, 100, or more million years? At first, it is reasonable to assume that things will continue in their present directions. The Atlantic will continue to widen, the Pacific will contract. The process which closed the Tethys Ocean will continue, with more earthquakes and mountain uplift in the hazardous country between the Alps and the Himalayas. Australia will continue north, catching on Borneo and twisting round to collide with China. Further into the future, some motions may reverse. We know that a predecessor of the Atlantic opened and closed in the past, and it is probably inevitable that the Atlantic Ocean crust will eventually cool, contract, and start to sink again, perhaps subducting under the east coast of the Americas. Then the continents will bunch up again. Christopher Scotese of the University of Texas in Arlington predicts that, 250 million years into the future, there will be a new super-continent, Pangaea Ultima, possibly with an inland sea, all that will remain of the once-mighty Atlantic Ocean.

Chapter 6
Volcanoes

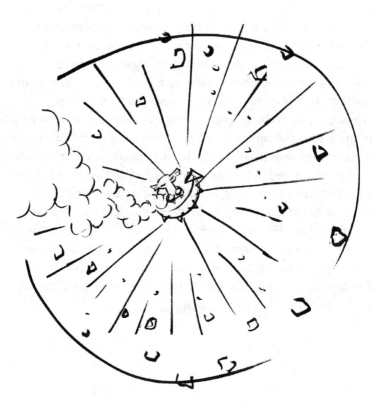

The relentless motions of tectonic plates, the uplift and the erosion of mountain ranges, and the evolution of living organisms are processes which can only be fully appreciated across the deep time of geology. But some of the processes at work in our planet can manifest all too suddenly, changing the landscape and destroying lives on a very human timescale: volcanoes. Superimpose a map of active volcanoes on a world map showing the boundaries of the tectonic plates and their association is obvious. The ring of fire around the Pacific, for example, is clearly associated with the plate boundaries. But where is the molten rock that feeds them coming from? Why are volcanoes different from each other, with some producing gentle eruptions and regular trickles of molten lava, whilst others erupt in devastating explosions? And why are some volcanoes, such as those of Hawaii, in the middle of the Pacific, far from any obvious plate boundary?

History is littered with eyewitness accounts of volcanic eruptions and with explanations, some of them mythological, some fanciful, and some surprisingly accurate. Among the better accounts is that of Pliny the Younger, describing the eruption of Vesuvius in 79 AD which killed his uncle, Pliny the Elder, and destroyed Pompeii and Herculaneum. But for a long time no one understood the cause of volcanic eruptions. Often they were thought to be the work of a fire god or goddess, such as the Hawaiian goddess Pelée. In medieval Europe, volcanoes were thought of as the chimneys of hell. Later, it was suggested that the Earth was a cooling star with remnants of the stellar fire within, linked through a system of fissures. In the 19th century, what we now know to be volcanic rock was widely thought to be deposited from oceans, the Neptunist theory, as opposed to the Plutonist idea that it had once been molten. After the Plutonist view gained ground, many thought that the interior of the Earth must be molten, an idea that did not get ruled out until the dawn of seismology. One of the mysteries was that volcanic rocks can have different compositions; sometimes even when erupted from the same volcano. Charles Darwin was one of the first to suggest that the composition of the melt could change as a result of dense minerals crystallizing out and sinking in the magma, something that he backed up by observations of volcanic rocks in the

Galapagos islands. As with his ideas on continental drift, it was Arthur Holmes in the mid-20th century who was the first to come close to the truth with his ideas of convection within the solid Earth's mantle.

How rocks melt

The key to understanding volcanoes comes from understanding how rocks melt. For a start, they don't have to melt completely, so the bulk of the mantle remains solid even though it gives rise to a fluid, molten magma. That means that the melt does not have the same composition as the bulk of the mantle. As long as the so-called dihedral angles, the angles at which the mineral grains in mantle rock meet, are large enough, the rock behaves like a porous sponge and the melt can be squeezed out. Calculations show how it will tend to flow together and rise quite rapidly in a sort of wave, producing lava at the surface in the sort of quantities seen in typical eruptions.

Melting does not necessarily involve increasing the temperature. It can result from decreasing the pressure. So a plume of hot, solid mantle material will begin to melt as it rises and the pressure upon it reduces. In the case of a mantle plume, that can happen at considerable depths. Helium isotope ratios in the basalt erupted on Hawaii suggest that it originates at around 150 kilometres depth. The mantle there is composed mainly of peridotite, rich in the mineral olivine. Compared to that, the magma that erupts contains less magnesium and more aluminium. It is estimated that as little as 4% of the rock melts to produce Hawaiian basalts.

Beneath the mid-ocean ridge system, the melting takes place at much shallower depths. Here there is little or no mantle lithosphere and the hot asthenosphere comes close to the surface. The lower pressures here can result in a larger proportion of the rock melting, perhaps 20 or 25%, supplying magma at about the right rate to sustain sea floor spreading and produce an ocean crust 7 kilometres thick. Most of the ocean ridge eruptions pass unnoticed as they take place more than 2,000 metres underwater as rapidly quenched pillow lavas. But seismic studies have revealed magma chambers a few kilometres beneath the sea floor in parts of the ridges, particularly in the Pacific and Indian oceans, though there is also some evidence of magma chambers beneath the mid-Atlantic ridge. Where a mantle plume coincides with an ocean ridge system, as in the case of Iceland, more magma is generated and the ocean crust is thicker, in this case rising above the sea to form Iceland.

Hawaii

The Big Island of Hawaii has welcoming people and friendly volcanoes. The town of Hilo is probably more at risk from tsunamis triggered by distant earthquakes than from the great 4,000-metre volcano of Mauna Loa that looms behind it. To the north and west lie the other Hawaiian islands and the Emperor seamount chain, tracing the long journey of the Pacific plate across the hot spot of an underlying mantle plume. To the south of the Big Island of Hawaii is Loihi, the newest of the Hawaiian volcanoes. As yet it has not broken the surface of the Pacific, but it has already built a high mountain of basalt on the ocean floor and will almost certainly become an island above water before long. Because Hawaiian lava is very fluid, it can spread over a wide area and does not tend to form very steep slopes. Such volcanoes are sometimes known as shield volcanoes, and they can flood basalt over a wide area. Often, a particular flow will develop a tunnel around it as the outer crust solidifies but the lava continues to flow inside. When the supply of lava ceases, the tunnel can drain and be left hollow.

The last big volcano to finish erupting on Hawaii, Mauna Kea, is home to an international astronomical observatory beneath some of the clearest skies on Earth. It was whilst visiting there one night that I saw, through binoculars, a distant fiery eruption on the flanks

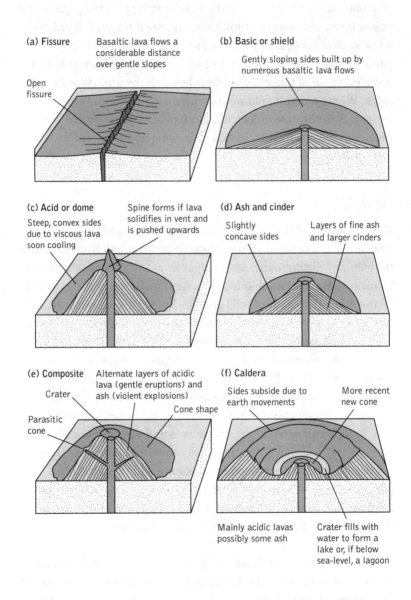

17. The principal types of volcano, classified by shape (not to scale).

of Mauna Loa, in the Puu Oo crater. The next day I was able to take a helicopter flight low over the freshly erupted lava flows. Through the open door I could feel the radiant heat from the glowing, and in places still moving, lava and smell traces of sulphur in the air. But it all seemed quite safe, even hovering in the crater itself, though avoiding the plume of smoke and steam. The nearby Kilauea caldera, where many of the current eruptions take place, boasts an observatory and viewing platforms. Every few weeks, visitors can witness a new eruption, often beginning with a curtain of fire with many fountains of glowing magma along a rift. Nothing is burning in the fire, but the release of volcanic gases through the hot, runny lava causes incandescent streams to fountain tens or even hundreds of metres into the sky. The eruption may last only a few hours. In spite of the high fountains, eruptions are not particularly explosive due to the very runny nature of the lava. This allows vulcanologists from the nearby Volcanoes Observatory to approach the molten lava, and even the vents, wearing heat-protecting suits. They will probably have already detected the magma rising in the vent through a sensitive network of seismometers and by measuring the change in the pull of gravity brought about by the upwelling magma. Sometimes they are able to collect uncontaminated samples of volcanic gas directly from the vents and take the temperature of the molten lava. It erupts at about 1,150 degrees Celsius.

Plinean eruptions

The nature of a volcanic eruption depends on the viscosity, or stickiness, of the magma and also on the amount of dissolved gas and water it contains. Early in an eruption any groundwater will be flashed explosively to steam. As the gas comes out of solution when the pressure is released, it will expand rapidly, sometimes explosively. Moderate amounts of gas in runny basalt produces the fire fountains of Hawaii. Greater quantities of gas will carry with it finely divided solid material such as ash and cinders. Eruptions tend to be more violent early on when the magma still contains a lot of gas. If it has had time to settle in a comparatively shallow magma chamber, it will be better behaved. Sometimes the gas and ash rise so high into the air that they spread widely before the ash falls. This was the sort of eruption witnessed on Vesuvius in 79 AD and has been called a Plinian eruption after the account by Pliny of his uncle's death.

Many different volcanic rocks can be produced in such eruptions. Where the ash and cinders are hard before they land, layers of loose tuff will build up. If the fragments are still molten, it would be a welded tuff. Near the vents, larger lumps of magma will be thrown out. If they are still molten when they hit the ground, they will form splatter bombs that look a bit like cow pats. If a solid crust forms in the air around the still-expanding lava bomb, it will form a breadcrust bomb, looking rather like a loaf of risen bread. Lava that is rapidly quenched can form volcanic glass called obsidian. If the lava solidifies with gas bubbles still in it, these are known as vesicles. Sometimes a foam of gas bubbles in lava can form, creating pumice of such low density that it floats on water. The surface of a lava flow can be very rough and cinder-like, in Hawaii called 'aa' lava. (This is a Hawaiian word, not just what you say when you try to walk over it!) Where a thin skin forms on a fluid lava flow, it can crinkle into flow lines, creating ropy or 'pahoehoe' lava. Occasionally, fine strands of lava become drawn out, an effect sometimes known as Pelée's hair.

Ring of fire

As long as you keep clear of the fire fountains and the fast-flowing lava, Hawaiian eruptions are reasonably safe. But that is not true of most volcanoes. Much of the Pacific is circled by a ring of fire – volcanoes of a much more temperamental nature. Where an ocean plate dives down in a subduction zone beneath a continent or an island arc, so-called stratovolcanoes develop. These can be picture-postcard volcanoes. Mount Fuji in Japan is one, with steep, conical slopes and a snow-capped peak topped by a smoking crater. But the beauty of such mountains can mask their sinister behaviour. They are notorious for earthquakes and sudden, violent eruptions, as Mount Unzen in Japan and Mount Pinatubo in the Philippines both showed in 1991. They are called stratovolcanoes because of their stratified structure of alternate layers of lava and ash or cinders, which can spew over an area far wider than the peak itself.

The reason for their violent behaviour compared to Hawaiian volcanoes is that they are not fed by clean, fresh magma from the mantle. The material sinking down beneath them is old ocean crust. It is saturated with water, both as liquid in pores and fissures but also bound into hydrous minerals. As the slab descends, it heats, due to its depth and also possibly due to friction. The presence of water lowers the melting point, so partial melting occurs. The pressure is so great that the water easily dissolves in the melt, lubricating it, and this magma squeezes up through the continental crust above. As it nears the surface, the pressure drops and the water starts to escape as steam. It can do so very rapidly and violently, in much the same way as gas escapes when a well-shaken bottle of champagne is uncorked.

On its way up, the magma may accumulate in chambers until there is enough pressure to erupt. During this time, dense minerals can solidify and fall to the floor of the chamber. These minerals, particularly iron compounds, are what make basalts dark and dense. The melt that remains for the eruption is lighter-coloured and richer in silica – up to 70 or 80% silica in some cases, compared to 50% or less in basalt. It forms rocks known as rhyolites and andesites, typical of places like Japan and the Andes. Their eruption is violent, not only due to the high water content but also because such silica-rich lavas are much more viscous and sticky. They do not flow easily, and bubbles can't easily escape. Such lava cannot fountain like a Hawaiian eruption but is flung out of the way explosively.

Mount St Helens

One of the most famous eruptions of recent times was of a volcano of this type. Mount St Helens is in Washington State in the northwestern USA, where the Pacific plate is subducting. At the beginning of 1980 it was a beautiful mountain surrounded by pine forests and lakes, a popular holiday retreat. It had shown little activity since 1857. Then, on 20 March 1980, in a series of small earth tremors building up to a magnitude 4.2 quake, it reawakened. Earth tremors continued to increase and trigger minor avalanches until 27 March, when a big explosion occurred in the summit crater and Mount St Helens began to spew ash and steam. The prevailing wind blew the darker ash to one side, leaving white snow on the other. The mountain took on a two-tone appearance.

So far, no lava had erupted. All that escaped from the crater was steam, blowing ash with it. But it signalled the rise of hot magma beneath the volcano. The earthquake activity continued, but the seismographs also began to record a continuous rhythmic ground-shaking different from the sharp shocks of the earthquakes. This so-called harmonic tremor is believed to have been generated by magma rising beneath the volcano. By mid-May, 10,000 earthquakes had been recorded and a prominent bulge had developed in the north flank of Mount St Helens. By firing laser beams at reflectors placed around it, geophysicists were able to measure its growth. It was pushing northwards at an incredible 1.5 metres per day. By 12 May, certain parts of the bulge were more than 138 metres higher than before the magma intrusion began. The volcano was literally being wedged apart into a highly unstable and dangerous condition.

Early on the morning of Sunday 18 May, Keith and Dorothy Stoffel were in a small plane above the mountain when they noticed rock and snow sliding inwards into the crater. Within seconds, the whole north side of the summit crater began to move. The bulge had collapsed in a great avalanche. It was like taking out the cork from a

18. The eruption of Mount St Helens in Washington State in 1980 was one of the most spectacular and best documented in recent times. The column of ash and smoke rose nearly 20 kilometres into the atmosphere.

champagne bottle. The magma inside it was exposed. The explosion was almost instantaneous. The Stoffels put their plane into a steep dive to gain speed and escape. David Johnston of the US Geological survey was not so lucky. An hour and a half before he had radioed in the latest laser beam measurements from his observation post 10 kilometres north of the volcano. As the north flank fell away, the blast headed straight for him. He was one of 57 people to die in the eruption.

Though it began several seconds later, the blast quickly overtook the avalanche. It fanned out at more than 1,000 kilometres per hour. Over a radius of about 12 kilometres, trees were not only flattened but swept away. Nothing living or man-made was left. As far as 30 kilometres away, trees were toppled, though isolated pockets survived in hollows. Even further afield, the leaves were seared by heat and branches were snapped.

Soon after the lateral blast, a vertical column of ash and steam began to rise. In less than 10 minutes it was 20 kilometres high and began to expand into a typical mushroom cloud. The swirling ash particles generated static electricity, and lightning started many forest fires. Wind soon spread the ash to the east, and space satellites were able to track it right around the Earth. Between 1 and 10 centimetres of ash fell over most of northwest USA. During the nine hours of vigorous eruption, about 540 million tonnes of ash fell over 57,000 square kilometres.

Yet another hazard of such eruptions are so-called pyroclastic flows. These are made up of particles of rock or magma shattered by explosions and swept along at several hundred kilometres per hour in a mass of hot gas. Their speed and temperature make them particularly deadly. When Mount Pelée in Martinique in the West Indies erupted in 1902, a pyroclastic flow swept down on the city of St Pierre, killing almost all the 30,000 inhabitants. Ironically, one of the two survivors lived because he was in solitary confinement in a thick-walled, poorly ventilated cell in the prison. Pyroclastic flows on Mount St Helens reached no further than the avalanche debris, though in some places where the material came to rest in old lakes, the heat was still sufficient to flash the water into steam and cause what looked like secondary eruptions. It may have been a pyroclastic flow that destroyed Pompeii in 79 AD.

Mount St Helens itself was left 400 metres lower than before the eruption, with a large new crater in the middle. There were several more explosive eruptions during 1988 and one in 1992, but none was as spectacular as the first. Today the mountain is bristling with scientific instruments that could pick up signs of further activity.

Blasts from the past

The eruption of Mount St Helens may seem tremendous, but it is small in comparison with others in the historical and prehistoric past. It threw 1.4 cubic kilometres of material into the air. The eruption of Tambora in Indonesia in 1815 by comparison ejected about 30 cubic kilometres, and that of Mount Mazama in Oregon in 5000 BC produced an estimated 40 cubic kilometres of ash. In 1883, the island of Krakatau (which is west of Java, not east as in the film title) blew up, leaving a 290-metre deep crater in the sea floor. Most of the 36,000 casualties were drowned in the resulting tsunami, in which a 40-metre wave stranded a steam ship deep in the jungle. Around 1627 BC it was the turn of the Aegean island of Santorini, or Thera, to blow. This happened at the height of the Minoan Bronze Age and may have contributed to the downfall of that civilization and the legend of the lost land of Atlantis. On the geological timescale, those are just the latest in a series of violent eruptions.

Anatomy of a volcano

Few volcanoes conform to the simple stereotype of a conical hill sitting above a tank-like magma chamber with lava pouring out of a summit crater. One of the most studied volcanoes is Etna in Sicily, and it is certainly more complex. It is a very active volcano and is probably only about a quarter of a million years old. But the rate of activity seen over the last 30 years cannot have continued for all that time or it would be even bigger. It's unlike Vesuvius and the island volcanoes of Vulcano and Stromboli to the north. They are stratovolcanoes fed from the subduction of the Ionian sea floor. Etna, by contrast, probably has its origins in a mantle plume. But its nature appears to be changing. Measurements of the composition of lava of different ages show that recently it began to take on more characteristics of the subduction-fed volcanoes to the north, and the nature of its eruptions do appear to be changing, becoming more violent and potentially dangerous.

The plumbing beneath a volcano such as Etna can be quite complex. There is not a system of hollow pipes awaiting the magma; it must force its way up by the route of least resistance. In a mantle plume, that preferred route is probably a lower-density column of material that can be easily pushed out of the way by the teardrop-shaped mass of rising magma. In the harder crust it must find a route through cracks and fissures. A big volcano is very heavy and can overload the crust it stands on, causing a network of concentric cracks. After the volcano ceases to be active it can collapse along such cracks, creating a wide caldera. While magma continues to rise it can force its way into the cracks, forming concentric swarms of conical sheets or ring dykes. The rise of magma within a volcano will make it bulge and crack in a series of minor earth tremors. In the case of Etna, the summit crater shows almost constant activity. I have made the steep climb up the loose cinder cone to peer in during a quiescent period. Even then, the ground feels warm to the touch and there is a smell of sulphur in the air. Gusts of steam rise from the vent, accompanied by a sound not unlike what I might imagine would be made by a snoring giant or a dragon.

Sometimes the dragon awakes and the crater rim is not a safe place to stand. As an eruption begins, blocks of hot rock up to a metre

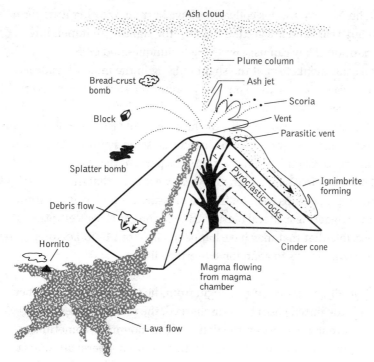

19. Some of the principal features of an erupting composite volcano such as Etna.

across can be thrown into the air. In 1979 an eruption started in this way but then fell silent after a period of heavy rain caused a slump inside the crater. However, the pressure built up and caused an explosion. Unfortunately, many tourists were standing around the crater rim at the time. Thirty were injured and nine were killed. Dr John Murray of the Open University recalls another occasion in 1986 when he was watching what appeared to be a fairly normal eruption with activity slowly building throughout the afternoon. The volcanic bombs were landing within 200 metres of the vent, well clear of the geologists. Then suddenly their range increased to more than 2 kilometres. Huge lumps of rock were whistling over the geologists' heads and landing around them. Statistically, the chances of being hit were not high but, as John Murray says, it did not feel like that at the time.

John Murray and his colleagues have been monitoring Etna for many years and are getting to know the signs of an impending eruption. They can use surveying techniques and GPS measurements to spot the slight bulging in the mountainside as magma rises beneath. They can monitor the earth tremors as cracks are forced open. Gravity surveys reveal the dense magma as it rises. The surveyors also monitor the hillside as an eruption subsides. In particular they are concerned about the steep southeast flank above the city of Catania. In the early 1980s parts of that flank subsided 1.4 metres in a single year, and there were fears that the slope might collapse, perhaps even taking the pressure off the magma within and triggering a lateral blast like the Mount St Helens eruption. It is possible that that may have happened in about 1500 BC, forcing the Ancient Greeks to abandon eastern Sicily.

Typically, an eruption will begin from the summit crater but then, once the initial gases have discharged, the magma forces a way out through fissures on the flanks of the mountain, sometimes dangerously close to villages. There have been several attempts to deflect the lava flows by excavating alternative channels, bulldozing up embankments, trying to stop the flow by hosing it with water or even blasting it. Sometimes homes have been saved, but sometimes they have not. In 1983 a flow stopped just touching the Sapienza Hotel in which the geologists stay. Human attempts to tame the volcanic forces still seem puny compared with the might of the mountain.

Volcanoes and people

Layers of ash and jagged lava flows can break down surprisingly quickly to yield a rich and fertile soil. People with short memories cluster around the volcano with their farms, villages, and even cities. It is possible to monitor volcanoes and get at least some warning when magma is rising within them and an eruption is imminent. But even then, it is sometimes difficult to persuade people to move. Where there is a very large population, such as

Major volcanoes

Name	Ht(m)	Major eruptions	Last
Bezymianny, USSR	2,800	1955–6	1984
El Chichón, Mexico	1,349	1982	1982
Erebus, Antarctica	4,023	1947, 1972	1986
Etna, Italy	3,236	Frequent	2002
Fuji, Japan	3,776	1707	1707
Galunggung, Java	2,180	1822, 1918	1982
Hekla, Iceland	1,491	1693, 1845, 1947–8, 1970	1981
Helgafell, Iceland	215	1973	1973
Katmai, Alaska	2,298	1912, 1920, 1921	1931
Kilauea, Hawaii	1,247	Frequent	1991
Klyuchevskoy, USSR	4,850	1700–1966, 1984	1985
Krakatoa, Sumatra	818	Frequent, esp. 1883	1980
La Soufrière, St Vincent	1,232	1718, 1812, 1902, 1971–2	1979
Lassen Peak, USA	3,186	1914–15	1921
Mauna Loa, Hawaii	4,172	Frequent	1984
Mayon, Philippines	2,462	1616, 1766, 1814, 1897, 1914	2001
Montserrat		Dormant until 1995	1995–8
Nyamuragira, Zaire	3,056	1921–38, 1971, 1980	1984
Paricutin, Mexico	3,188	1943–52	1952
Mount Pelée, Martinique	1,397	1902, 1929–32	1932
Pinatubo, Philippines	1,462	1391, 1991	1991
Popocatepetl, Mexico	5,483	1920	1943
Mount Rainier, USA	4,392	1st century BC, 1820	1882
Ruapehu, New Zealand	2,796	1945, 1953, 1969, 1975	1986
Mount St Helens, USA	2,549	Frequent, esp. 1980	1987
Santorini, Greece	1,315	Frequent, esp. 1470 BC	1950
Stromboli, Italy	931	Frequent	2002
Surtsey, Iceland	174	1963–7	1967
Unzen, Japan	1,360	1360, 1791	1991
Vesuvius, Italy	1,289	Frequent, esp. 79 AD	1944

around the Bay of Naples below Vesuvius, it may not be practical or possible to evacuate people in time. Less well-known volcanoes such as many of those in South America have not even had impact studies performed on them.

Slow-moving lava flows and the more deadly pyroclastic flows are not the only dangers. The clouds of dust and water vapour that can be built around an erupting volcano can produce heavy rain and this, perhaps coupled with melting snow on the volcano, can lead to catastrophic mud flows or lahars, like the one that swept down the flanks of Nevado del Ruiz in Colombia in 1985 killing about 22,000 people. The threat can even be invisible. Dissolved volcanic carbon dioxide accumulated in deep waters in Lake Nyos in Cameroon. One cold night in 1986 a build-up of dense, cold water at the surface suddenly sank, bringing gas-rich water to the surface and releasing the pressure on it. It was like opening a well-shaken bottle of soda, leading to a sudden release of the gas, which, being heavier than air, swept down a valley and suffocated 1,700 people as they slept in their villages.

Volcanic forces may be unstoppable, but with prudent planning and careful monitoring, people can learn to live with them in comparative safety.

Chapter 7
When the ground shakes

A supertanker crossing the ocean under full steam carries a lot of momentum. Its stopping distance will be many kilometres. An entire continent will stop for nothing. We've already heard about the 55-million-year slow-motion collision of India with Asia. The other tectonic plates are also moving with respect to one another. Where they grate against each other they make earthquakes. A map of major earthquakes shows up the tectonic boundaries even more clearly than volcanoes do.

GPS measurements show how the tectonic plates are slowly and steadily gliding past one another at several centimetres per year.

But as you approach the plate margins, the motions become less smooth. There are places where the movement is steady, without major earthquakes, as if the rocks were lubricated or so soft that they can move by the mechanism known as creep. But many of the plate boundaries get stuck. The continents keep moving, however, and strain builds up until the rocks can take it no longer and suddenly crack, producing an earthquake.

Some earthquakes occur at great depths as ocean crust subducts into the mantle. But most quakes happen in the top 15 to 20 kilometres, where the crust is hot and brittle. The rocks break along what are known as fault lines, sending out seismic waves. The waves appear to radiate out from a focus or hypocentre underground along the fault. The point on the ground surface above the hypocentre is called the epicentre.

Earthquake magnitudes

The Richter and Mercalli scales plot the magnitude of earthquakes. The former measures the actual wave energy, while the latter charts the destructive effects. The Richter scale is logarithmic, so, just using the numbers one to ten, it can accommodate everything from the frequent daily tremors that go by almost unnoticed in a seismically active area to the largest earthquakes recorded – to date, the largest has been on the coast of Chile in 1960, which measured 9.5 on the scale. The difference in energy between each point on the scale is a factor of 30. So, for example, a factor 7 quake is likely to be much more destructive than a factor 6. Ironically, many of the personal records of Charles Richter, the Californian seismologist who gave his name to the scale and who died in 1985, were destroyed in a house fire following the magnitude 6.6 Northridge earthquake near Los Angeles in 1994.

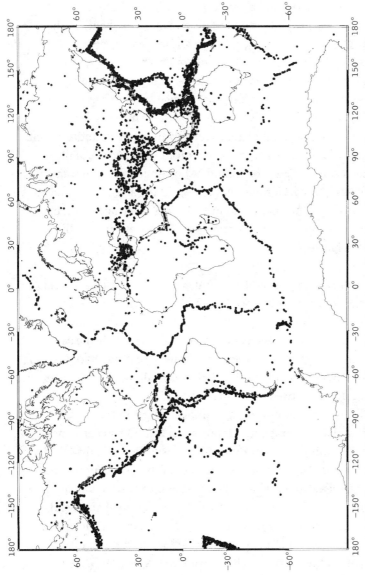

20. The distribution of major crustal earthquakes (magnitude 5 and above) in the past 30 years. Most cluster along tectonic plate boundaries, but a few occur mid-continent.

The most famous crack in the world

In California earthquakes are almost a regular feature of life. The great Pacific plate is on the move, not diving beneath the American continent but grating past it in what is called a strike-slip fault. The junction is seldom a straight line, so kinks in the main fault line result in swarms of many parallel and intersecting cracks or faults. Most of them suffer frequent minor earthquakes and any of them could be the centre of a big one. The most famous fault of all, effectively the plate boundary itself, is the San Andreas fault. It can be traced from the south of California, curving round inland of Los Angeles and running northwest straight for San Francisco and the sea. It achieved its notoriety in 1906 when San Francisco was devastated by a major earthquake and the terrible fire that followed it through the wooden houses.

Between Los Angeles and San Francisco, the landscape is arid and the fault can be followed easily through the bare hills. Sometimes it is marked just by a slight change in the slope. Sometimes it can be seen cutting through the landscape as though some great hand had run a knife across the map. It seems to run straight for 100 miles. I followed it along a rough farm track midway between Los Angeles and San Francisco. To the east lie low eroded hills of the Temblor range; to the west, the dry Carrizo plain slopes gently towards San Luis Obispo and the Pacific. Coming down from the hills are a number of dried-up stream beds. As they reached the base of the slope, something strange seems to have happened to them. Instead of flowing straight on to the west, they all take a sharp right-hand bend through 90 degrees, follow the base of the hills north for a few tens of metres, and then turn sharp left again to continue to the sea. One of the biggest of them, Wallace Creek, named after Robert Wallace of the US Geological Survey, is cut deep into the soft hillside. As it crosses the fault, it is displaced by 130 metres. Originally it must have flowed straight down the slope, cutting its course. In a series of earthquakes, the plain to the west has lurched north, taking the stream bed with it. The winter floods could not cut a new course through the high banks, so they followed the fault until they found their old bed again. It did not move all at once. Using a combination of excavation and radio-carbon dating, Robert Wallace and his colleagues have worked out the stages. The only one recorded in the history books took place in 1857 and accounts for the last 9.5 metres of slip. Its two immediate predecessors, both of them prehistoric, moved the stream courses a further 12.5 and 11 metres. Averaged out, the San Andreas fault has been slipping at a rate of 34 millimetres per year for the past 13,000 years. If it keeps up this rate, in 20 million years' time Los Angeles will be as far north of San Francisco as it is south today. But, as Californians know to their cost, the passage is not smooth.

Measuring the movement

It used to be almost impossible to measure motions of metres or centimetres across the scale of continents, but now it is comparatively easy. Fault zones such as California and Japan bristle with instruments. In particular, receivers linked to the GPS can keep a continuous tab on their position on the surface of the globe. If they are linked into an automatic monitoring network, they can inform the authorities immediately exactly where a quake has taken place and how severe it is. As we shall see, they may also help to give advance warning. An even clearer image of exactly what has happened can come from space. Remote-sensing satellites equipped with synthetic aperture radar, or SAR, can record the shape of the ground surface so accurately that when two images are superimposed, one taken before, the other after a quake, they produce interference patterns which reveal the precise section of the fault which moved and its motion.

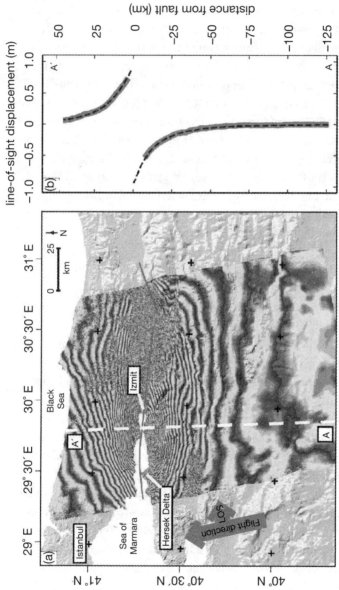

21. Satellite radar interference map combining data from before and after the Izmit quake in Turkey in 1999 to reveal the ground movement.

Mid-plate quakes

Even the seemingly rigid slabs of the continents are subject to stresses and strains, and occasionally they move. The biggest earthquake in the brief recorded history of the United States took place not in California but in the eastern USA. In 1811 the frontier town of New Madrid near St Louis was rocked by three massive quakes measuring up to 8.5 on the Richter scale. They were powerful enough to ring church bells in Boston and, had great modern cities existed then in the broad Mississippi plain, they could have been flattened. No one is sure whether the quake was due to the ground subsiding under the weight of Mississippi sediments or whether the mighty river itself owes its existence to the crust being stretched. It could be that this is another line along which an ocean tried to open, before it chose the Atlantic the other side of the Appalachian Mountains. Perhaps it is having another try. Whatever its cause, another New Madrid earthquake today could cause untold destruction.

The mystery of deep earthquakes

By plotting the depths of earthquakes, it is possible to trace the descent of ocean lithosphere in a subduction zone such as the one where the Pacific plate descends beneath the Andes of South America. For the first 200 kilometres or so, the rock is cold and brittle so fractures and generates earthquakes as it would nearer the surface. But some earthquakes seem to have their foci far deeper, up to 600 kilometres down, where heat and pressure should make the rock soft and ductile so that it deforms rather than fractures. A possible explanation is that these deep earthquakes might be due to a whole layer of crystals undergoing a phase change from the olivine structure found in ocean lithosphere to the denser spinel structure of the mantle. One argument against this theory is that this process can only happen once, yet several earthquakes have been recorded from about the same place. But maybe it is just successive layers of olivine transforming.

Awaiting the inevitable

In January 2001, northwestern India was rocked by a devastating quake centred on the town of Bhuj in Gujerat. This was part of the continuing legacy of the intercontinental collision of India with Asia. The relative motion between India and Tibet still adds up to about 2 metres per century. Though there were a number of severe Himalayan quakes during the 20th century, there are many areas that must have accumulated far more strain. A slip of 2 metres has the potential to produce a magnitude 7.8 earthquake. But there are some parts of the thrust where India is pushing under the Himalayas that have accumulated a strain equivalent to a slip of double that. In fact, some areas have not experienced a severe quake for more than 500 years. Such a 'great quake' could be devastating indeed. Although building standards have improved in the past century, the evidence from Bhuj suggests that a similar magnitude earthquake now is likely to kill a similar proportion of any population as it could have done 100 years ago. But in the meantime, populations at risk have increased by factors of 10 or more. If the 1905 Kangra quake were repeated today, 200,000 fatalities are quite likely. Should one of the major cities in the Ganges plain be hit, the figure might be worse by yet another order of magnitude. Another highly populous earthquake zone, Tokyo, has not had a major quake since 1923. If a great quake were to strike there today, even with Japan's improved building standards, it could cause an estimated $US 7 trillion of damage, which might lead to the collapse of the global economy.

Designing for earthquakes

It is often said that it's buildings, not earthquakes, that kill people. Certainly, most casualties in quakes result from collapsed buildings and the subsequent fires. Many factors affect whether a building will fall down in an earthquake or not. Clearly the strength of the quake is important, but so is how long the shaking continues. Then there is the design of the building. For small structures, bendy or

22. Raised motorways have been spectacular casualties of recent urban quakes. This is Kobe City in Japan in 1995.

compliant materials can be better than hard, brittle ones. Just as trees can sway in the wind, so a wood-framed building can sway in a quake without falling down. Lightweight structures are less likely to kill people when they do fall. But the traditional wood and paper partitions in Japanese homes are more prone to fire. The worst buildings in earthquakes are probably those of brick or stone and ones with frames of poor-quality reinforced concrete; just the sort found throughout the poorer earthquake-prone countries. The 1988 quake near Spitak in Armenia and the 1989 Loma Prieta one near San Francisco were both magnitude 7, but the first killed more than 100,000 people, whereas only 62 died in the second, largely because of the very strict building regulations

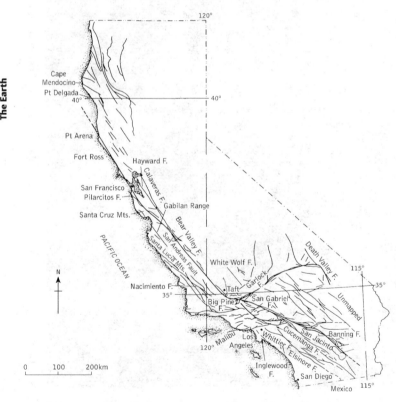

23. The San Andreas fault system in California is not a single crack in the Earth's crust. Even this map is simplified.

in California. Tall buildings there are very strong and built so that they do not resonate at the frequencies of earthquake waves. Many have rubber blocks in their foundations to absorb the vibrations. In Japan, some skyscrapers have systems of heavy weights in the roof that can be moved quickly to cancel out the shaking due to a quake.

When the ground turns to liquid

If you've ever stood on a wet sandy beach and jiggled your foot up and down, you may have noticed that water tends to rise in the sand and your foot sinks in; the sand liquefies. The same thing happens when an earthquake shakes wet sediments. The epicentre of the Mexico City earthquake in 1985 was 400 kilometres away, yet many buildings in the city were destroyed. They were built on land reclaimed from an ancient lake and the earthquake waves resonated to and fro through the mud for nearly three minutes, liquefying it so that it could no longer support buildings. However deep the foundations of buildings are, they don't provide much support if the ground turns to liquid. In both the 1906 and 1989 quakes near San Francisco, the worst-damaged buildings were in the Marina district, built on reclaimed land.

Fire

One of the greatest dangers when an earthquake strikes a city is fire. Both in San Francisco in 1906 and Tokyo in 1923, more people lost their lives in fires than in the quakes themselves. Fires can start easily as cooking stoves are overturned, and spread quickly in the tangled ruins of wooden buildings, fuelled by fractured gas pipes. In San Francisco, the fire services were inadequate; fire engines were trapped in their garages or by blocked streets, and fractured water mains drained the city's water supply. Today, quake-prone cities such as San Francisco are developing systems of so-called 'smart pipes', both for gas and water, that will quickly and automatically shut off sections of pipe where the pressure falls suddenly due to a break.

Saving lives

The safest place to be during an earthquake is in open, flat countryside. Just about the worst thing to do is to panic. In a city, falling glass and masonry outside can be more of a hazard than staying indoors under strong structures such as stairways. Schoolchildren in Japan and California have regular training in how to protect themselves. Yet, in real earthquake situations, most people still tend either to freeze where they are or panic and run into the open. If people are caught in a collapsed building, there's a whole range of heat-sensitive cameras and listening devices to find them, and every disaster brings tales of miraculous rescues as well as tragedy.

Chance and chaos

In one way, earthquakes can be predicted with certainty. Cities such as San Francisco, Tokyo, and Mexico City will definitely experience another earthquake. But that knowledge is not much use to those who live there. They want to know precisely when the Big One will come and how severe it will be. But that is just what geologists cannot tell them. Like the weather, the Earth is a complex system in which a tiny cause can have a big effect. Like the imaginary Amazonian butterfly that can supposedly influence the weather in Europe by flapping its wings, so a pebble stuck in a fault could lead to an earthquake. Although it may never be possible to predict earthquakes with certainty, predictions in terms of probability are getting better and better; the closer to the time of the quake, the better the accuracy of the forecast.

Traditional signs

Long before the age of scientific instruments, people had been looking for early warnings of an impending quake. The Chinese in particular have become quite adept at noticing strange animal behaviour, sudden changes of water level and gas content in wells, and other signs that could prelude a quake. Using such indicators, the City of Haicheng was evacuated in 1975, hours before a devastating earthquake, saving hundreds of thousands of lives. But, a year later 240,000 people died in Tengshan, where no warning had been given. Other clues may include tiny flashes of light and electricity, possibly produced as mineral crystals are squeezed, in the same way that squeezing a piezo-electric gas lighter produces a spark. There is serious research into how animals are able to sense an imminent quake; for example, in Japan to see if catfish may behave abnormally due to electrical disturbances. But what constitutes abnormal behaviour in a catfish? And how many householders will monitor one? There's also some evidence that a big quake may be preceded by very low-frequency electromagnetic waves. But the best indicators seem to be in the pattern of seismic waves running through the ground.

Playing the odds

Most big earthquakes are preceded by foreshocks. The trouble is, it's hard to say whether a minor earth tremor is an isolated event or the prelude to a major quake. But it can change the odds. From historical records, it may be possible to say that a big quake is likely sometime in the next 100 years. But that puts the chance of an earthquake happening tomorrow at one in 36,500. There maybe ten minor tremors a year, any of which could be the foreshock heralding a big quake. So the detection of a minor tremor increases the probability of the quake in the next 24 hours to one in 1,000. By understanding where all the faults are, when they last cracked, and by having instruments in all the right places, it is sometimes possible to increase the accuracy of the prediction to one in 20. But

Major earthquakes

Location	Year	Magnitude	Deaths
Bhuj, India	2001	7.7	20,085
El Salvador	2001	7.7	844
Peru	2001	8.4	75
Taiwan, China	1999	7.7	2,400
Turkey	1999	7.6	17,118
Afghanistan	1998	6.1	4,000
N Iran	1997	7.1	1,560
Russia (Sakhalin)	1995	7.5	2,000
Japan (Kobe)	1995	7.2	6,310
S California	1994	6.8	60
S India (Osmanabad)	1993	6.4	9,748
Philippines	1990	7.7	1,653
NW Iran	1990	7.5	36,000
San Francisco (Loma Prieta)	1989	7.1	62
Armenia	1988	7.0	100,000
Mexico City	1985	8.1	7,200
N Yemen	1982	6.0	2,800
S Italy	1980	7.2	4,500
NE Iran	1978	7.7	25,000
Tangshan, China	1976	8.2	242,000
Guatemala City	1976	7.5	22,778
Peru	1970	7.7	66,000
NE Iran	1968	7.4	11,600
Nan-shan, China	1927	8.3	200,000
Japan	1923	8.3	143,000
Gansu, China	1920	8.6	180,000
Messina, Italy	1908	7.5	120,000
San Francisco	1906	8.3	500
Calcutta, India	1737	-	300,000
Hokkaido, Japan	1730	-	137,000
Shensi, China	1556	-	830,000
Antioch, Turkey	526	-	250,000

that is still the same as saying that there is a 95% chance that there *won't* be a quake tomorrow – hardly a statistic to announce over the radio and evacuate a city. It may, however, be sufficient to alert emergency services and to stop the transport of hazardous chemicals.

Real time warnings

Predicting earthquakes in advance may always be difficult. But you can detect them with certainty when they happen. This can be turned into a sort of early-warning system. This was tested in California in 1989 following the Loma Prieta earthquake. An elevated section of the Nimitz freeway had partially collapsed and rescue workers were trying to free motorists trapped underneath. The huge concrete slabs were unstable and aftershocks could have sent them crashing down. The focus of the earthquake was almost 100 kilometres away, so sensors on the fault itself could radio a warning at the speed of light which would reach the rescuers 25 seconds ahead of the shock waves themselves, travelling at the speed of sound, giving people time to scramble clear. In the future, such a system could be used to give a brief warning of the main shock of an earthquake. For example, shock waves from the main fault lines behind Los Angeles would take up to a minute to reach the city. A radioed warning would not be enough for evacuation but, linked to computer systems, it could help banks to save their accounts, elevators to come to a standstill and open their doors, automatic valves to seal off pipelines, and emergency vehicles to get clear of buildings.

Epilogue

This introduction to a wonderful planet has indeed been short. I have endeavoured to give an overview of some of the key processes at work, beneath our feet and above our heads. I have tried to show how these dynamic processes interact at the surface to give the world we know and love. From those processes grow a rich diversity of land forms, rocks, and living communities. I have not attempted to introduce the beautiful minerals and crystals that make up the rocks of our planet. I have not investigated the details of the processes through which those rocks are tossed by tectonic forces and carved by wind, rain, and ice into the often breathtakingly beautiful landscape in which we live. I have not looked at how the ground-down remains of those rocks are deposited in sediments nor how they produce fertile soils on which our food chain depends. Neither have I looked at the most fantastic product of all of our planet – life – and how the physical forces of our world have conspired with chemistry and natural selection to make ours a living planet. All these things are worthy of more complete introductions of their own.

I do, however, believe that our planet is very special, and that, without its rare combination of geophysical processes, life, at least as we know it, would not have had the chances it has. I have tried to show how everything is interdependent. Without water, rocks would not be lubricated, granite might not form, and we would not have the great land masses of continents. Without water, there would be no clouds and no rain; just a wind-blown desert landscape with little possibility for life. Without liquid water, the chemistry of life could not function and life as we know it could not exist on Earth. Without life, there would not be the feedback mechanisms on atmospheric composition that have, so far at least, kept the climate bearable. Without life, the Earth might now be a snowball world or a super-heated greenhouse.

In spite of things being just right for life for most of the last billion years, we are still at the mercy of our planet. The tectonic forces of volcanoes and earthquakes are more than a match for the atmospheric forces of floods, droughts, and storms. Both destroy the lives and livelihoods of millions. Yet somehow we survive. For the most part, we scurry about like ants on the surface, oblivious to the larger picture. But even so, humans have become a powerful force themselves in shaping the planet. Through urbanization and agriculture, civil engineering and pollution, we have transformed a large fraction of the land surface. It has been at a price. The rate of extinction of animal and plant species may be greater today even than during the mass extinctions at the end of the Cretaceous and Permian periods. The atmospheric composition – and, as a result, the climate – seem to be changing faster now than at any time since the last Ice Age and possibly for a lot longer.

We are no longer the victims of our planet, we are the custodians of it. Through our inconsiderate greed for land and our disregard for pollution, we bite the hand that feeds us. But we do so at our own peril. We still have all our eggs in one basket, all our people on one planet. We need to care for that planet and take responsibility for it. But we also need to progress with the search for new homes and the technology to take us to the stars.

Regardless of what we do to it, our world will not last forever. At any moment it could be decimated by a collision with an asteroid or comet. Or it could be spit-roasted by radiation from a nearby exploding star. And we could produce much the same effect by engaging in all-out nuclear war even sooner. Ultimately, in about 5 billion years' time, the Sun will run out of hydrogen fuel in its core and start to expand into a red giant star. The latest estimates suggest that the incandescent gas will not reach quite as far as the Earth, though it will certainly engulf Mercury and Venus. But it will scorch our beautiful world to a cinder, driving off oceans and atmosphere, and rendering it uninhabitable. But 5 billion years is a long time even for a planet. As species go, the human race is statistically unlikely to survive 5 million, let alone 5 billion, years. Maybe new dominant life forms will emerge on Earth. Maybe we will evolve or engineer ourselves into something different. Maybe our descendants will find a way of encapsulating memories and consciousness into immortal machines. Overall, I am an optimist and like to imagine planetary scientists of the future exploring and colonizing new worlds and comparing them with the dynamic planet we call Earth.

Further reading

T. H. van Andel, *New Views on an Old Planet* (Cambridge University Press, 1994). A good overview of plate tectonics and our dynamic world.

P. Cattermole and P. Moore, *The Story of the Earth* (Cambridge University Press, 1985). An astronomical perspective on our planet.

P. Cloud, *Oasis in Space* (W. W. Norton, 1988). Earth history from the beginning.

G. B. Dalrymple, *The Age of the Earth* (Stanford University Press, 1991)

S. Drury, *Stepping Stones* (Oxford University Press, 1999). The development of our planet as home to life.

I. G. Gass, P. J. Smith, and R. C. L. Wilson, *Understanding the Earth* (Artemis/Open University Press, 1970 and subsequent editions). This OUP introductory text has become a classic.

A. Hallam, *A Revolution in the Earth Sciences* (Clarendon Press, 1973)

P. L. Hancock, B. J. Skinner, and D. L. Dineley, *The Oxford Companion to the Earth* (Oxford University Press, 2000). An encyclopaedic reference from hundreds of expert contributors.

S. Lamb and D. Sington, *Earth Story* (BBC, 1998). A readable account of Earth history based on the TV series.

M. Levy and M. Salvadori, *Why the Earth Quakes* (W. W. Norton, 1995). The story of earthquakes and volcanoes.

W. McGuire, *A Guide to the End of the World* (Oxford University Press, 2002). A catalogue of catastrophes, real or potential, that could strike our planet. Not for the fearful!

H. W. Menard, *Ocean of Truth* (Princeton University Press, 1995). A personal history of global tectonics.

R. Muir-Wood, *The Dark Side of the Earth* (George Allen and Unwin, 1985). A history of the people involved in making geology into a 'whole Earth' science.

M. Redfern, *The Kingfisher Book of Planet Earth* (Kingfisher, 1999). A lavishly illustrated introduction for younger minds.

D. Steel, *Target Earth* (Time Life Books, 2000). The role of cosmic impacts in shaping our planet and threatening our future.

E. J. Tarbuck and F. K. Lutgens, *Earth Sciences*, 8th edn. (Prentice Hall, 1997). Another classic text.

S. Winchester, *The Map that Changed the World* (Viking, 2001). How William Smith published the first geological map in 1815.

E. Zebrowski, *The Last Days of St Pierre* (Rutgers University Press, 2002). Fascinating historical account of the geological and human factors that led to the volcanic destruction of St Pierre in Martinique.